BASIC ANATOMY OF A CAT

Ear canal

Third eyelid

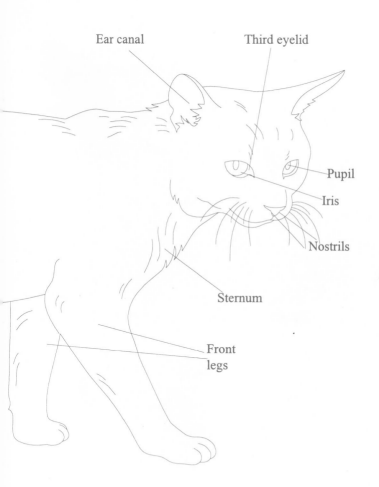

Pupil

Iris

Nostrils

Sternum

Front
legs

what your
cat
is trying to
tell you

Produced by The Philip Lief Group, Inc.

ST. MARTIN'S GRIFFIN
NEW YORK

A Head-to-Tail Guide
to Your Cat's Symptoms—
and Their Solutions

what your
cat
is trying to
tell you

John Simon, DVM
with
Stephanie Pedersen

Special Note to Readers

This book was written with the goal in mind of helping pet owners identify with their cat's health problems. It furthermore provides the pet owner with a general idea of what to expect when they take their sick cat to their own veterinarian for diagnosis and treatment. Finally, home-care tips that can be used until professional veterinary care is obtained are discussed.

Although Dr. Simon practices both conventional and alternative medicine at his own clinic, this books limits his discussion to the conventional approach to diagnosis, and treatment as it would normally occur at most veterinary clinics. Certain recommendations for homecare using antioxidants, mega vitamin-mineral therapy, and a few herbs are the exception to this rule.

WHAT YOUR CAT IS TRYING TO TELL YOU. Copyright © 1998 by The Philip Lief Group, Inc. All rights reserved. Printed in the United States of America. No part of this book may be used or reproduced in any manner whatsoever without written permission except in the case of brief quotations embodied in critical articles or reviews. For information, address St. Martin's Press, 175 Fifth Avenue, New York, N.Y. 10010.

Production Editor: David Stanford Burr

Library of Congress Cataloging-in-Publication Data

Simon, John, D.V.M.
       What your cat is trying to tell you : a head-to-tail guide to your cat's symptoms and their solutions / John Simon, with Stephanie Pedersen.—1st ed.
          p.   cm.
       Includes index. V 5-11-98 769428
       ISBN 0-312-18213-9
       1. Cats—Health. 2. Cats—Diseases. I. Pedersen, Stephanie.
II. Title.
SF985.S55   1998
636.8'0896075—dc21   97-40590
                                              CIP

First St. Martin's Griffin Edition: March 1998

10 9 8 7 6 5 4 3 2 1

To my wife Joanie,
my office manager Judy,
and my entire hospital staff without whose daily
support and encouragement I could never have
found the time or energy to complete this book.

Contents

CHAPTER 3
Head and Neck • 21

CHAPTER 4
Eyes, Ears, and Nose • 35

CHAPTER 5
Mouth and Throat • 53

CHAPTER 6
Hair and Skin • 61

CHAPTER 7
Chest, Heart, and Lungs • 74

CHAPTER 8
Abdomen • 95

CHAPTER 9
Spine, Limbs, and Paws • 136

CHAPTER 10
Tail and Anus • 147

APPENDIX A
Checklist for Good Health • 152

APPENDIX B
How to Perform a Weekly Home Exam • 160

APPENDIX C
Breed Disease Predilections • 163

APPENDIX D
Important Questions to Answer Before
Going in for an Exam • 168

APPENDIX E
List of Recommended Dosages • 169

Index • 193

About the Authors • 207

Introduction

Cats are not like people when it comes to pain: They can't vocalize where they hurt, hazard a guess about what ails them, or explain the extent of their suffering. What they *can* do is give you certain physical signals that something could be wrong. Yet, how do you decode these clues to determine whether a condition is serious or not? At the first sign of something out-of-the-ordinary, should you rush to the vet in a panic and spend a good sum only to discover that nothing is wrong? Or is this the one time your hunch might be correct—and that immediate response is the only course of action that will save your loved one's life?

When something doesn't seem right with your pet, you need quick, reliable answers you can easily understand. Enter *What Your Cat Is Trying to Tell You: A Head-to-Tail Guide to Your Cat's Symptoms and Their Solutions*, an easy-to-follow medical reference written strictly for pet owners—not veterinarians. This is an important distinction; it means this book is designed expressly for those who have no knowledge of veterinary medicine but have a strong desire to help their beloved companions.

What Your Cat Is Trying to Tell You focuses on symptoms, those clearly observable tips-offs that something isn't right. To use this book, simply note your cat's symptoms, check the Contents to find your cat's most prominent symptom, and follow the page number to the corresponding section in the book. Find the entry that contains the combination of related symptoms that best match those your kitty displays. Pointed questions about your pet's habits, age, sex, breed, and health history help guide you. Your answers act as clues that uncover what illness your pet may have, and also help you deduce how your cat became ill. Accompanying the descriptive rundown of symptoms is advice on care and important information that can help you prevent the situation in the future. The extensive index also may help you locate more specific information.

Each of the chapters concerns a different region of the body. The book opens with a chapter on emergency symptoms—impossible-to-ignore signs that something is seriously wrong. Here is where you'll receive guidance on how to handle crises.

Chapter 2 covers behavior. Is your kitty acting strangely—mewing excessively, urinating on the rug, or lashing out at the slightest provocation? This chapter can shed light on undesirable conduct and help you determine whether an illness or an environmental factor is to blame. Of course, you'll also get advice on addressing the predicament.

Chapters 3–10 address symptoms confined to specific body areas, including runny eyes in Chapter 4 (Eyes, Ears, and Nose) and loss of appetite in Chapter 8 (Abdomen).

Finally, convenient appendixes put essential information at your fingertips. Here you'll learn how to perform a weekly home exam. You'll find a good-health guide that helps you determine your pet's grooming and health-maintenance needs. There's even an appendix listing feline breeds and the illnesses to which each may be prone.

In many cases, you'll definitely want to seek the advice of a professional veterinarian. And if you're ever in doubt about whether or not to call the vet, you'd be wise to call and check. But whether you're waiting to hear back from the vet or simply searching for ways to prevent future problems, *What Your Cat Is Trying to Tell You* offers easy-to-follow techniques to understanding your best friend's needs.

CHAPTER 1

Emergency Symptoms

Perhaps you've actually witnessed your pet meet with misfortune, such as in a vehicle run-in, a fight with another cat, or taking a few quick sips of a poisonous chemical before you could stop him. Or maybe you've come across him after such a catastrophe. Either way, you're probably familiar with many emergency symptoms. After all, common sense is all you need to realize that if your kitty exhibits strong, dramatic signals—such as gushing blood, an exposed bone protruding from his thigh, an obvious wound, extremely heavy vomiting that doesn't seem to stop, or perhaps even unconsciousness—you need to sprint to the vet.

Yet, what about those subtler signs of trouble? The ones that don't announce themselves, that maybe even show up gradually, but are every bit the emergency warning signs their more dramatic counterparts are? Sure it takes an adroit owner—one who is thoroughly versed in her cat's normal behavior and vital signs—to spot anything unseemly, but most pet owners are just such people.

The point of all this is that just because the sign you happen to notice is subtle, it's still worth taking seriously. Indeed, some such signs, such as a change in posture, an unwillingness to move, a high fever, and an unusually low body temperature, can indicate life-threatening conditions. You should also be aware that because felines generally prefer to lick their wounds—literally—in private, a cat that won't come out from under the bed or has recently gone missing may be injured or seriously ill. Conversely, other, more showy symptoms, such as coughing, sneezing, or loud wheezing, usually indicate nothing more grave than a kitty cold or a feline allergy.

Abdomen: Acute Pain

You can tell your pet's abdomen hurts if he shrinks from being touched there, moves cautiously or not at all, or adopts the prayer pose with his chest against the floor and rump in the air. Other signs of abdominal pain include refusing food, trembling, vomiting, extreme restlessness, inability to find a comfortable position, irregular purring, grunting, meowing and crying, and labored breathing. If the condition comes on suddenly, it commonly signals an emergency situation. **Get your pet to the vet immediately**. [Also see sections in Chapter 8, Bloated, Distended, or Painful Abdomen (p. 96); and Obvious Bulge at Midabdomen, Groin or Rectal Area (p. 96).]

Bite and Scratch Wounds

Being the territorial creatures they are, fights between outdoor felines are surprisingly common. If you suspect your cat has had a row with a neighborhood puss (disheveled fur and bald patches are common giveaways), give your pet a thorough once-over, paying special attention to his ears, eyelids, and the base of his tail. Check for bite and scratch wounds.

Minor scratches can be swabbed with an antiseptic, such as 3% hydrogen peroxide. If the wound appears deep, wrap the area with a pressure bandage to slow the bleeding. If left untreated, such a wound can cause considerable blood loss. Now you're ready to go straight to the vet. Often embedded in such wounds are dirt, saliva, and hair from the rival cat—all of which can cause the wound to become an infected, abscessed mess. These contaminants can also cause blood poisoning. To prevent an infection, your vet will remove any embedded dirt and hair, flush the wound with 3% hydrogen peroxide, clean it with a surgical soap, and then rinse it with water. Finally, she will apply a wound antibiotic and bandage the area. Antibiotic injections may also be given and oral antibiotics dispensed.

Bleeding: Uncontrollable

Maybe your cat was cut by a piece of glass. Maybe he was hit by a vehicle. Maybe he snagged himself jumping over or squeezing through a fence. In any event, your kitty is bleeding and you want to make it stop.

Bleeding comes in degrees of seriousness. A relatively minor cut will typically stop bleeding by itself after 6 or 7 minutes. A larger cut may not stop bleeding without assistance from you. Find a *clean* rag, towel, or gauze bandage and hold it directly against the wound. (Here's an optional step designed to keep hair out of the wound—especially important if your pet is a longhaired cat: Before applying the cloth, smear a thin coat of petroleum jelly around the outside of the cut.) If the blood flow stops, you can hold off seeing the vet for 4 or 5 hours (but you'd be wise to have it checked at that point in the event that the wound may need cleansing and/or suturing). If the flow doesn't stop in 20 minutes, **you must go straight to the vet.** After the vet has halted the blood flow, she may stitch the area closed, providing it is not badly contaminated and can be easily cleaned.

You also must go straight to the vet anytime there is a deep cut in the cat's chest area. On the way to the doctor's office, apply a pressure bandage to the area. Should a rib protrude through the skin, avoid touching the bone with your hand or with the bandage. With a chest wound, you may notice air bubbles in the leaking blood. You may also hear a hissing noise: This sound is actually air escaping from the chest cavity.

If, for some reason, you can't get a wound to stop bleeding heavily after a 30 minutes worth of pressure, and you cannot get to a vet immediately—perhaps you live in a rural area—you may need to apply a tourniquet. Tourniquets are used *only* on appendages—in other words, the cat's limbs and tail.

To apply a tourniquet, find a soft, elastic fabric, such as a sock. Tightly tie the fabric around the appendage, directly *above* the wound. To avoid killing living skin and muscle tissue, you will have to loosen the tourniquet every 10 minutes, for 30 to 45 seconds at a time, to allow blood to flow into the limb. Once the blood flow has significantly slowed—or stopped altogether—replace the tourniquet with a pressure bandage.

Break, Fracture, or Sprain: Difficulty Moving

A fall or a collision with a car or bike can break or fracture a bone or injure ligaments. First, if you do see your kitty hit by a vehicle or involved in a serious fall, it's a good idea to take the animal to a vet straightaway for a checkup. (Read the third paragraph of this section for information on how to transport an injured cat.)

The more likely scenario, however, is that the cat was hurt during an unsupervised moment. Thus, you may suddenly notice your kitty limping (if a leg was injured) or adopting a hunched-up posture (if a rib was cracked or the spine was damaged). You may also notice a cracking or grinding noise when your pet moves, as well as swelling at the injured site. You may even see a piece of bone protruding through the skin.

If you do see a piece of errant bone, you can be certain there's a break. Without visible bone peeking through, however, determining whether the cat has a fracture or a sprained or torn ligament can be difficult for nonvets. Regardless of the actual problem, encourage your kitty to remain still while you *gently* pick him up (using both hands) and lay him on a blanket with the injured area facing *up*. Be aware that a cat in pain may bite indiscriminately out of fear and discomfort. If another person is present, each of you should take an end of the blanket and lift your pet—hammock style—into an appropriately sized, stable box, or other open, sturdy container (if there is no one to help, you'll have to do this yourself). Moving him as little as possible, take your pet to the vet, who will radiograph the injured limb, set a broken or fractured bone, wrap a sprain, or surgically repair torn ligaments.

Breathing: Extremely Difficult or Stopped; Unconsciousness

A breathing problem caused by a respiratory condition is not an emergency. However, the following are cause for immediate action: your kitty has consumed poison, had a run-in with a car or bicycle, or had a serious fall and is aggressively struggling for breath or has quit breathing altogether and lapsed into unconsciousness.

If the cat is still conscious, do not try to examine his mouth—you are likely to get bitten. **Get to the vet immediately.** If the cat is unconscious, ask someone to help you (if possible) so you can administer CPR (cardiopulmonary resuscitation) en route to the vet.

Start by opening his mouth and lying the tongue to one side between the top and bottom molars. Place your finger in the cat's mouth and throat to feel for any obstructions, including vomit, mucous, or blood. Remove anything you find. *Extend the cat's head and neck, then close his mouth. Inhale deeply, completely cover your pet's nose and mouth with your mouth, then exhale into his nostrils.* The air should reach his chest: Watch for the chest to swell. Remove your mouth and allow the cat's chest to deflate normally. When it has, put your mouth over his nose and mouth and start again. This inflate-deflate cycle should be done 12 times per minute, until the cat begins breathing on his own.

Often, immediately after a cat has ceased breathing (or just prior to it), his heart may stop. (You can check for his heartbeat by wrapping your hand around his chest, just behind his front legs, and applying slight pressure. Alternately, you can check the cat's pulse, which is best felt on the inner side of the thigh in the groin area.) If the nearest vet is some distance away, you also will have to perform external cardiac compression. Place the cat on his right side, laying him on the firmest surface possible. Place your thumb on the side of his sternum (chest) that is facing up and wrap your other four fingers around the chest so that your fingers are underneath his body and pressing on the other side of the sternum. Squeeze the chest firmly between the thumb and four fingers. Release. Then squeeze firmly again. Release. Repeat 6 times, then wait 5 seconds to see if the chest expands and whether any pulse or heartbeat results. If it doesn't, repeat. When you combine pulmonary resuscitation with external heart massage, you should perform 1 pulmonary expansion for every 6 cardiac compressions.

Burns

Cats, being curious creatures, may find themselves face-to-face with spitting cooking grease, scalding-hot water, fire, scorching surfaces, or even a caustic chemical substance, such as lye. If nearby, you'll hear your kitty's pained meow, know that he's been burned, and can immediately treat the injury. Because it's entirely possible that your kitty was burned while out of your sight, here's what to look for: Immediately after being burned, the area will be red and painful. If the burn is bad, the skin may even turn a shade of white or brown. The skin will be blistered and may appear shriveled. If fire or a hot surface caused the injury, fur may be singed or missing altogether. Note: After being injured, most cats tend to hide—under beds, in closets, in bags. Should your cat act reclusive, there is a real chance that he is injured. Approach him tenderly and gently examine him.

Cool the burn by holding an ice pack or bag of frozen vegetables against the wound for 15 minutes. If you suspect a caustic material is to blame, rinse the area with cold water and diluted shampoo—it will help rid the fur and skin of any remaining chemicals. For acid burns, neutralize by rinsing with 1 teaspoon of baking soda per pint (2 cups) of water. For alkali burns, neutralize by rinsing with 2 tablespoons of vinegar per pint (2 cups) of water. Cover the wound with a clean, dry cloth or a gauze bandage and *go straight to the vet.*

Convulsions: Continuous

Your pet begins moving in a disjointed, jerky way. Suddenly, seizures take over his body—not just one seizure, but one right after the other with no letup. Your cat may (or may not) slip into unconsciousness, vomit, or lose bowel and bladder control. What should you do? Gently place a towel or thin blanket over the animal and **head straight for the nearest veterinary clinic**—even if it's not the one you go to regularly. Such nonstop seizures often lead to exhaustion, unconsciousness, then death.

En route to the vet, do not place your hand anywhere near your kitty's mouth—you can be severely bitten. And don't worry about whether your convulsing cat will swallow his tongue—it rarely happens. If you do suspect poisoning, be aware that you *should never induce vomiting when convulsions are present.*

Cats who experience only 1 or 2 convulsions in 24 hours still should see a vet in order to get the condition diagnosed, but the condition is not an emergency unless a single seizure lasts for more than 10 minutes. While observing a cat having a seizure, make sure the poor animal does not injure himself against the furniture or any sharp object. Pillows work well to protect both pet and owner.

Drowning

Most cats have no great love of water, yet they can swim short distances. But if they fall into your backyard swimming pool, they can't climb over the pool's edge, which means they could easily drown if there is no nearby exit ramp for them to use. Cats have also drowned when hiding in an open washing machine that the unsuspecting owner closes and turns on.

Should you see your pet floating in a pool, bathtub, or other expanse of water, remove him immediately. At this point, he may be weak or even unconscious. *Quickly turn him upside down, suspending him by his back legs.* This lets any inhaled water exit his windpipe. Next, place the cat with his head lower than his chest—you can use pillows or rolled up clothing or towels to make an artificial slope or you can place his body on a sofa or bed with his head positioned over the edge—and begin artificial respiration.

If the cat is completely unconscious, you will need to administer CPR. Start by opening his mouth and lying the tongue to one side, between the top and bottom molars. Place your hand in his mouth and throat to feel for any obstructions, including vomit, mucous, or blood. Remove anything you find. Extend the cat's head and neck, and close his mouth. *Inhale deeply, completely cover your pet's nose and mouth with your mouth, then exhale into his nostrils.* The air should reach his chest: Watch for the chest to swell. Remove your mouth and allow the cat's chest to deflate normally. When it has, put your mouth over his nose and mouth and start again. This inflate-deflate cycle should be done 12 times per minute, until the cat begins breathing on his own.

Often, immediately after a cat has ceased breathing (or just prior to it), his heart may stop. This means you may also have to perform external cardiac compression. Place the cat on his right side, laying him on the firmest surface possible. Place your thumb on the side of his sternum (chest) that is facing up and wrap your four fingers around the chest so that your fingers are underneath his body and pressing on the other side of the sternum. Squeeze the chest firmly between your thumb and fingers. Release. Then squeeze firmly again. Release. Repeat 6 times, then wait 5 seconds to see if the chest expands and whether a pulse or heartbeat results. If it doesn't, repeat. When you combine external heart massage with pulmonary resuscitation, you should produce 1 pulmonary expansion for every 6 cardiac compressions.

Once your kitty has begun breathing on his own, take him to the vet, since many cats who get water in their airways develop pneumonia.

Electrocution

Kittens and untrained cats love to play with toys if you provide them, but they will just as happily occupy themselves with things they find around the house, including shoes, furniture, and electrical cords. Although most of these objects won't harm your cat, chewing on electrical cords is a quick way for him to become electrocuted. (Note: You can try to keep cats from chewing cords by coating the surface with a bitter-tasting substance—such as Bitter Green or Bitter Apple, available at your pet store—formulated expressly for this purpose.) The safest bet is to keep cords unplugged whenever possible. Less-common causes of electrocution include coming in contact with power lines, being struck by lightning, and touching exposed wires.

Typically, your cat doesn't chew on cords when you're around to catch him, which means you may enter a room to find your kitty lying on the floor unconscious, cord in his mouth. As long as the animal still has the cord in his mouth, he's probably still being shocked. **Do not touch the cat or you will be electrocuted, too!** Instead, immediately switch off the electrical source via the circuit breaker. If, for some reason, it is impossible to shut off the electricity, use a nonmetal object, such as a stick, broom handle, or wooden chair, to move the cat away from the cord. If there is any water or urine on the floor, push your cat away from the liquid.

Another possible scenario may involve a cord lying next to your convulsing pet, who happens to be conscious. To distinguish electrocution from an epileptic fit, look for pale burns around the mouth, surrounded by red, swollen tissue.

Check your cat for vital signs. If the cat is still conscious, do not try to examine his mouth—you are likely to get bitten. **Get to the vet immediately**. If the cat is unconscious and not breathing, ask someone to help you (if possible) so you can administer CPR en route to the vet.

Start by opening his mouth and lying the tongue to one side, between the top and bottom molars. Place your hand in the mouth and throat to feel for any obstructions, including vomit, mucous, or blood. Remove anything you find. Extend the cat's head and neck, and close his mouth. *Inhale deeply, completely cover your pet's nose and mouth with your mouth, then exhale into his nostrils.* The air should reach his chest: Watch for the chest to rise. Remove your mouth and allow the cat's chest to deflate normally. When it has, put your mouth over his nose and mouth and start again. This inflate-deflate cycle should be done 12 times per minute, until the cat begins breathing on his own.

Often, immediately after a cat has ceased breathing (or just prior to it), his heart may stop. If the nearest vet is some distance away, you will also have to perform external cardiac compression. Place the cat on his right side, laying him on the firmest surface possible. Place your thumb on the side of his sternum (chest) that is facing up and wrap your other four fingers around the chest so that your fingers are underneath his body and pressing on the other side of the sternum. Squeeze the chest firmly. Release. Then squeeze firmly again. Release. Repeat 6 times, then wait 5 seconds to see if the chest expands. If it doesn't, repeat.

If your cat isn't breathing and his heartbeat has ceased, the two procedures must be combined. Alternate 1 artificial respiration with 6 heart compressions until reaching the vet—or until the cat's breathing and heartbeat kick in.

Fever

A cat's normal temperature ranges from 100°F to 102.5°F. Your pet's temperature, however, can rise to 103.5°F if he has been exercising recently, if he is scared, nervous or excited, or if the air temperature is high. If your kitty's temperature is higher than that—especially if he has had a fever for more than 1 day, his body is indicating that his immune system is responding to an internal threat or imbalance. Depending on what other symptoms your pet exhibits—anything from coughing to lethargy, vomiting to confusion—he could have some type of infection, endocrine disorder, cancer, drug reaction, tissue inflammation, or autoimmune disease. **You should bring your cat to the vet.**

Heat Stroke

Just like humans, animals who linger in hot environments for too long are susceptible to heat stroke. Warning signs include excessive panting, physical collapse, a body temperature as high as 108°F, and unconsciousness.

You can attempt to cool your pet by immersing him in a tub of cool water or giving him a cold sponge bath. Continue covering your cat with cool water until his temperature is below 104°F, **then rush your pet to the vet.** She will give your kitty medication and possibly cold-water enemas to further lower the body's temperature.

You can prevent heat stroke by never leaving your pet shut up in a hot car or garage, and by making sure that he never hides in the dryer.

Hypothermia, Overexposure to the Cold

Considering the way cats seek out warm windowsills and fresh-from-the-dryer laundry, perhaps it's no surprise that they fare better in warm weather than in cold. In fact, if left in a cold temperature for too long, they can quickly lose body heat and develop hypothermia. This is especially likely when the cat is wet, for example, if you let him outside during a snowstorm.

Signs of hypothermia include shivering, lethargy, and a body temperature of 97°F or below. Warm your kitty by wrapping him in a blanket, coat, or towel. If he happens to be wet, immerse him in a warm bath, remove him, and rub him vigorously with absorbent towels to remove moisture from his fur. If your kitty will let you, use a hairdryer set on "low" to dry him.

If your kitty isn't wet, you can raise his body temperature without bathing him by applying warm-water packs to his armpits, chest, and abdomen (these can be hot-water bottles or washcloths moistened with very warm water—not hot enough to burn the skin). As the packs lose their heat, you'll have to replace them with fresh, warm ones until your pet's rectal temperature reaches 100°F. As your pet begins to feel better, he will become more active.

Frostbite often can accompany hypothermia. Obvious targets include ears, paws, and tail. If any of these areas appear "burned," treat your cat for hypothermia first, then sponge the frostbitten areas with warm water, bundle the cat tightly, **and take him to the vet.**

Insect Bites

Because of the amount of time cats typically spend outdoors, they are prone to bites and stings from a wide variety of insects, including ants, wasps, spiders, and bees. It's a good sign your pet has been zapped by something if there is a swollen area on his skin or if he suddenly becomes weak (due to an allergic reaction to the bite). If weakness is present, **take your cat immediately to the vet**—he may be having an abnormal or allergic reaction. Also rush to the vet's office if your feline's eye is swollen shut or if an entire side of his face appears distorted. These are classic symptoms of a poisonous spider bite.

If nothing else seems amiss, you can soothe your pet's pain by applying a paste (made of baking soda and just enough water to create a pasty consistency) directly to the afflicted area. For more information about bites from ticks carrying Lyme disease, see section in Chapter 3, Swollen Lymph Nodes, Fever, and Neurological Abnormalities, pp. 32–33.

Shock

Shock is the failure of the cardiovascular system to provide the body tissue with oxygen. It almost always accompanies another serious condition, such as poisoning, physical injury, heavy blood loss, or a severe allergic reaction. Shock can even appear after the cat has been weakened by a lengthy illness. Regardless of what causes shock, the result is the same: unconsciousness and eventual death.

Shock is a symptom umbrella, encompassing signals like quietness, inactivity, and nonresponsiveness to external environmental stimuli. Should any combination of these be present, take your cat's pulse, which is best felt on the inner side of his thigh in the groin area. A cat in shock will have a rapid pulse that grows weaker as shock progresses. His heart and respiratory rates will also be rapid. Due to a lowered body temperature, legs and paws will grow cool to the touch. The gums may appear very red or pale, depending on the cause of shock.

If possible, press your finger against your kitty's gum. The area will blanch beneath your finger. In healthy cats, the area will regain its normal color in 1 to 2 seconds. If a cat is in shock, the color will take longer to return.

However, you don't want to waste valuable time trying to diagnose your cat's condition. If you suspect he is in shock, bundle your pet in a warm blanket to preserve whatever body heat he has left and **immediately go to the nearest vet.**

Vomiting/Diarrhea: Repeated Due to Poisoning and Illness

If your cat eats an unfamiliar food or garbage, then vomits 2 or 3 times during 1 hour or struggles with diarrhea, you probably do *not* have an emergency on your hands—unless his abdomen is tense and painful or the frequency continues. If moderate vomiting and/or diarrhea has only been going on for 1 day, withhold all food for 24 hours and limit drinking to frequent but small drinks of Gatorade, sugar water (3 parts water to 1

13

part maple syrup), or uncarbonated Coke (made flat by shaking out the bubbles). A bland, cooked hamburger and rice diet can be served after the 24-hour fast.

You do have a crisis when the animal vomits and/or has bouts of diarrhea continually for more than 24 hours, or if the cat vomits blood, salivates excessively, collapses, loses consciousness, breathes rapidly, or has pale gums or large pupils. These symptoms commonly indicate poisoning or advanced stages of a stomach condition, such as gastritis, bloat, stomach ulcers, stomach tumors, foreign bodies, and intestinal obstruction (see Chapter 8). No matter what has prompted the symptoms, **you must get your cat to a vet immediately** for diagnosis and treatment. Dehydration can quickly result from repetitive vomiting and/or diarrhea, so wait no more than 24 hours before getting help.

CHAPTER 2

Abnormal Behaviors

There are two types of abnormal behavior in cats: that which suddenly manifests itself in your pet, and that which may come naturally to a cat but is undesirable to us human types. Both are important to address. Because there's always an underlying reason for a sudden shift of feline conduct, the first scenario represents a change in the physical or mental wellness of your pet that, for health's sake, should be examined. A common example of just such an abnormal behavior is when a litter-trained cat suddenly uses the living room rug as her personal restroom.

As for the second type of abnormal behavior, though it may seem like we humans are imposing our wills on our feline friends, it's important to keep in check many behaviors that may come naturally to our particular pets. Take aggression, for instance. Your kitty has to live in a world of humans, and if she lunges at, scratches, or bites a person, her existence in the land of homo sapiens is not going to be a pleasant one.

Aggression

The term aggression refers to numerous behaviors—all of them undesirable: excessive yowling; chasing, lunging at, and/or hissing at humans and other animals; biting or scratching; and destroying property. An aggressive cat often exhibits a combination of these.

Before labeling your cat as dangerous, however, putting her behavior in the proper perspective is essential. For instance, a young cat—especially one with no feline housemates—often hones her hunting techniques on her owner's legs, hands, and feet. From a kitten's viewpoint, this makes absolute sense. Although she probably has toys to play with, they are inanimate, and frankly, pose no kind of challenge. Her owner's body, on the other hand, represents a living, constantly moving target. If you happen to be the beleaguered owner, your only recourse is to provide your pet with "substitute prey" on which she can sharpen her skills.

Most cats have favorite nooks and crannies from where they wait to pounce. When passing one of these, carry with you some type of string toy you can wriggle, thus diverting her attention away from your flesh. Also, keep in mind that young cats need a lot of playtime. If your kitten has no feline companions, you will need to spend 10 to 30 minutes playing with her in the morning and evening. The only thing that mellows a playful kitten is exhaustion.

When looking at aggressive behavior, you also have to keep in mind your cat's reproductive ability: In other words, is your cat fixed? As an unneutered kitten reaches adulthood—and sexual maturation—she can become territorial and more dominant and protective. Male cats have earned a well-deserved reputation in this arena; however, unspayed females can be just as forceful—often yowling excessively when in estrus (in heat), or hissing and spitting when anything (real or imagined) threatens her newborn offspring. Neutering your animal is the most effective way to curb such behavior.

If the aggression comes on suddenly, take a look at your homelife. Have you made any changes lately? Taken a job that keeps you away from home more? Have you moved? Have you had a baby or visitors with children? Do you have a new roommate—or even a weekend houseguest? Have you introduced a new pet into your home, such as another cat, a dog, or a rabbit? Cats can be jealous and they are creatures of habit—they don't like their routines upset or their territory invaded. Aggressive behavior can be their way of expressing just how uncomfortable they are with the new arrangement. In the wake of any changes, spending extra time with your kitty is often a good start. Your vet can offer good advice on how to deal with such problems.

If nothing has changed at home, illness may be the culprit. Sudden aggression can be brought on by pain, by rabies, or by other conditions affecting the brain, such as brain tumors. Whatever the cause—sexual maturity, a change in environment, or a medical problem—your vet should be consulted. Aggressive cats are dangerous. They can attack other cats, neighborhood children—and even their owners.

Ailment: Burns •SYMPTOMS: Skin appears red, inflamed, or blistered. Animal may bite, lick, or paw at injured area. •FIRST AID: *Thermal burns:* apply ice or cold-water soaks for 20 minutes; *chemical burns:* flush area with large amounts of cold water; *acid burns:* neutralize with ¼ teaspoon of baking soda per pint (2 cups) of water; *alkaline burns:* neutralize with 2 tablespoons of vinegar per pint of water. Bring your cat to the veterinarian.

Ailment: Choking •SYMPTOMS: Gagging and drooling; swallowing is painful, so appetite often diminishes. •FIRST AID: While someone restrains the animal, carefully check for any objects in his throat. Don't probe too deeply: Additional gagging may worsen the condition. Bring your kitty to the vet.

Ailment: Cold Over-Exposure •SYMPTOMS: Unconsciousness; the body is cool to the touch, and the cat's temperature is below 98°F. •FIRST AID: Apply warm water-packs to the armpits, chest, and abdomen. Give the cat a warm bath and rub him vigorously. Continue this until his temperature is greater than 100°F. See your veterinarian as soon as possible.

Ailment: Cuts and Hemorrhaging •SYMPTOMS: A moist, open, bloody wound. The cat may lick the area continually. •FIRST AID: Clip hair around the wound and clean the surrounding area with antibacterial cleanser. Use 3% hydrogen peroxide to flush the wound itself. If the bleeding is on the limb and is profuse, apply an elastic bandage (or clean sock) as a tourniquet *above* the wound; release for 30 to 45 seconds every 10 minutes, and then reapply. Contrary to popular opinion, a cat who licks a wound actually hinders healing and introduces bacteria. Get the animal to a veterinarian promptly.

Ailment: Dehydration •SYMPTOMS: The cat is usually listless and weak. The skin becomes less elastic and the mouth becomes dry, with a dull appearance to the eyes and coat. Severe dehydration commonly results from prolonged or frequent vomiting and/or diarrhea. •FIRST AID: If your cat vomits 2 to 3 times in one day, try this stopgap measure: Place in the cat's mouth a small amount of Gatorade (½ to 1 ounce of fluid—depending on the estimated degree of dehydration—per pound of body weight); repeat this frequently. If vomiting continues for more than 24 hours, or if within that time period it occurs more than once every few hours, see your veterinarian as soon as possible—dehydration is potentially fatal.

Ailment: Electrocution •SYMPTOMS: The cat may be unconscious or convulsing on the floor. Pale burns may be seen around the mouth, surrounded by red, swollen tissue. You may notice an electrical cord in the cat's mouth or on the floor nearby. •FIRST AID: DO NOT touch the cat if he still has the cord in his mouth or you will be shocked, too. Turn the electricity off (via the circuit breaker). If that is not possible, move the cat away from the cord with a stick or nonmetal object. Check the vital signs, and if the animal is unconscious, use CPR to revive him. Start by opening his mouth and lying the tongue to one side between the top and bottom molars. Place your finger in the cat's mouth and throat to feel for any obstructions, including vomit, mucous, or blood. Remove anything you find. Extend the cat's head and neck, then close his mouth. Inhale deeply, completely cover your pet's nose and mouth with your mouth, then exhale into his nostrils. The air should reach his chest: Watch for the chest to swell. Remove your mouth and allow the cat's chest to deflate normally. When it has, put your mouth over his nose and mouth and start again. This inflate-deflate cycle should be done 12 times per minute, until the cat begins breathing on his own.

Often, immediately after a cat has ceased breathing (or just prior to it), his heart may stop. (You can check for his heartbeat by wrapping your hand around his chest, just behind his front legs, and applying slight pressure. Alternately, you can check the cat's pulse, which is best felt on the inner side of the thigh in the groin area.) If the nearest vet is some distance away, you also will have to perform external cardiac compression. Place the cat on his right side, laying him on the firmest surface possible. Place your thumbs on the side of his sternum (chest) that is facing up and wrap your other four fingers around the chest so that your fingers are underneath his body and pressing on the other side of the sternum. Squeeze the chest firmly between the thumb and four fingers. Release. Then squeeze firmly again. Release. Repeat 6 times, then wait 5 seconds to see if the chest expands and whether any pulse or heartbeat results. If it doesn't, repeat. When you combine pulmonary resuscitation with external heart massage, you should perform 1 pulmonary expansion for every 6 cardiac compressions. Get the cat to the vet. If the cat's heart stops, you may need to perform external cardiac compression on the way to the vet's.

Ailment: Eye Infections •SYMPTOMS: The eye appears red, swollen, or inflamed. The area around the cat's eye is matted with mucous and discharge.

• **FIRST AID:** Clean the matted area around the eye with warm water. Check for injuries or penetrating wounds in the eye itself. Since the eye is extremely delicate, get immediate professional care and leave all treatment of the eye itself to the vet (unless he tells you otherwise).

Ailment: Fits and Convulsions • **SYMPTOMS:** The cat runs frantically and "swims" on the floor (uncontrolled running action while lying on his side). His eyes appear glassy, and the cat may foam at the mouth. The jaws often clench and the teeth may grind. The cat's neck is arched backward and his limbs will be held in rigid extension. Most seizures last from 30 seconds to 5 minutes. • **FIRST AID:** Be careful! Wear gloves and heavy clothing for protection and move the cat to a dark, quiet room. Keep him from overheating; place ice packs on his head if he can be handled. Use pillows to keep him from injuring himself or you. Rush your cat to the vet.

Ailment: Fracture • **SYMPTOMS:** The leg dangles loosely. The cat limps and any movement may be accompanied by severe pain. A hairline or greenstick fracture (a break in which the bone is bent but broken only on the outside of the bend) may not be so obvious. • **FIRST AID:** Wrap the cat in a blanket for safe handling. Place him on plywood so that he can be carried without undue movement. Keep him warm with a blanket, and rush to the veterinarian.

Ailment: Heat Stroke • **SYMPTOMS:** Excessive panting, physical collapse, and unconsciousness. The cat's body temperature may be as high as 108°F. • **FIRST AID:** Cool the cat by immersing him in a tub of cold water, hosing him down, or giving him a cold sponge bath. Continue until his temperature is below 104°F. Get your cat to your vet as soon as possible.

Ailment: Ingested Objects • **SYMPTOMS:** Swallowed bone fragments, glass, or other foreign objects may bring on recognizable symptoms: The cat cries when picked up, passes a bloody stool, vomits, and may be constipated. • **FIRST AID:** Feed your pet bread or cereal: Food particles may cover sharp edges of the object to prevent further internal damage. Bring your cat to the vet.

Ailment: Insect Bites • **SYMPTOMS:** Swelling in the area of the bite. The cat may be weak if he suffers an allergic reaction to the bite. • **FIRST AID:** Apply baking-soda paste (made by adding just enough water to create a pasty consistency) directly to the sting. If weakness develops, see your veterinarian immediately. Spider bites can cause an eye to become swollen shut or the whole side of the face to appear distorted. Bee bites can do the same.

Ailment: Motion Sickness • **SYMPTOMS:** The cat may drool and/or vomit. • **FIRST AID:** Stop the car and calm the cat or take him for a short walk, providing you have a secure harness or leash.

Ailment: Overexertion • **SYMPTOMS:** The cat is stiff, moans, and has a hard time finding a comfortable resting position. • **FIRST AID:** Take your cat for an easy walk. Antioxidants such as vitamins C and E may be helpful in reducing stiffness and inflammation.

Ailment: Overindulgence • **SYMPTOMS:** The cat has diarrhea and is bloated, restless, and gassy. This may follow a night of garbage raiding or begging (meowing) at the table during holiday get-togethers. • **FIRST AID:** Place 1 teaspoon (or 5 eyedropperfuls) of a nonlaxative antacid (such as Maalox) directly into his mouth. Tap him on the nose or blow in his nose to get him to swallow. (Activated charcoal, available at drugstores, helps absorb gas and toxins.)

Ailment: Painful Abdomen • **SYMPTOMS:** The abdomen is extremely tender to the touch, and is usually accompanied by vomiting, extreme restlessness, meowing, and labored breathing. **FIRST AID:** Do not attempt to treat the cat yourself. Acute abdominal pain can signal a serious emergency that must be handled promptly by a veterinarian.

Ailment: Poisoning • **SYMPTOMS:** The cat cries, crouches, drools, and may suffer intense abdominal pain. He may also go into a coma. Symptoms are extremely varied and depend on the type of poison consumed. • **FIRST AID:** Read the warning label on the package and follow recommendations. As a general rule, induce vomiting by placing 1 teaspoon of salt on the back of your pet's tongue. Do not induce vomiting if the poison is an acid, alkali, solvent, heavy-duty cleaner, petroleum product, or tranquilizer. In this case, **rush your cat to the vet.**

Ailment: Shock • **SYMPTOMS:** Weakness, panting, and hard breathing; the cat's extremities may be cold to the touch. Shock is often associated with severe injury, and could result in a coma. • **FIRST AID:** Keep the cat warm and quiet. Cover him with a blanket or a heating pad. If breathing has stopped, pass an ammonia solution under his nostrils—the sneeze reflex may restart breathing. Rush to the vet's office.

Ailment: Skunk Odor • **SYMPTOMS:** The cat's coat emits a strong, foul odor. • **FIRST AID:** Bathe your pet in a scented cat shampoo and tomato juice. If his eyes are inflamed, seek veterinary care. A product called Skunk Off, available at most pet stores, also may prove helpful.

Ailment: Split Nails • **SYMPTOMS:** Bleeding from the foot may occur, and the cat may limp. • **FIRST AID:** Wash the injured toe with antibacterial cleanser. Wrap it with cotton and adhesive tape and bring him to the veterinarian.

Apprehension/Timidness

Cats are cautious by nature. Whereas dogs may bound over—tails wagging—to strangers in anticipation of making new friends, cats prefer a cooler approach. After watching a newcomer from afar, a cat may creep over to sniff a shoe or pant leg, or—if it's offered—the palm of a hand. After examining the person, she may decide to make contact, or she may walk away nonplussed. This does not make her shy, only indifferent.

Yet some cats *are* especially timid. New situations make them hiss with fear: a visitor sends them running for the safety of another room; a Christmas tree or new piece of furniture prompts them to walk yards out of their way to avoid it. They may even display an exaggerated aversion to pickup trucks, brooms, or some other object. Perhaps the kitty was born that way, or perhaps the animal was abused or otherwise traumatized before you adopted her. Either way, interacting with such a cat in a soothing, kind, patient manner can help maintain her calm. If your kitten shies from a particular object, you might be able to counteract the situation. Introduce the object to your cat gradually—this may mean once every other day for a year—and reward your cat with affection or a treat when she doesn't run from it.

In an otherwise moderately outgoing or extroverted cat, apprehension typically signals illness. When something hurts—maybe due to a stomachache caused by gastritis, an ear infection, or even a dislocated limb—an animal will physically shrink from you, fearing you may try to touch the painful spot. A normally gregarious cat who hides or wants to be left alone is letting you know that she is not feeling right. Take your pet to the vet for a thorough checkup.

Inappropriate Elimination Habits

Cats do not need to be trained to use the litter box. In all but the rarest cases, even the youngest kittens only need to be placed in the litter box to know what it is for. This is because it's a built-in instinct for felines to scratch in earthy material, eliminate, scratch, and cover up this material on their own. When a cat or kitten doesn't use the provided litter box, you can be sure there's a reason. If you consider how clean cats are, it's not surprising that the most common reason a cat shuns her litter box is that it needs to be cleaned. When was the last time you emptied the box, washed it, and supplied fresh litter? Try to scoop away waste products at least 2 times a day, and wash the box and provide a 2-inch layer of fresh litter 2 times a week—more often if the cat is sick and the box is seeing heavy use. (Of course, in a home with more than one cat, the litter box should be cleaned and refilled more frequently.) Another, easily remedied reason your pet may decide to use your floor rather than her box: You've switched to a brand of litter she doesn't like. Switch back to the old litter and she should do fine.

Still more litter-box trivia: Your cat values her privacy and may not use a litter box situated in a well-trafficked area, nor one placed near her food or bed. Try putting her box in a dry corner of your bathroom—and remember to keep the room's door open so she can access it whenever she needs to.

If none of these explain why your cat has suddenly taken to using your Persian rug as her toilet, the culprit may be psychological. Stories of cats "forgetting to use their litter box" after undergoing some owner-inflicted offense are numerous. Have you done anything recently to upset your cat? This can be anything that changes her routine, from a new job that reduces the playtime the two of you used to enjoy, to a new pooch or kitten that incites jealous tension, to a move to a new home. Punishment rarely works on cats. Instead, try giving your pet extra attention and 1 to 2 weeks to grow accustomed to whatever new arrangement has upset her in the first place. If, after 14 days, the litter box remains unused, it's time to visit the vet—both for behavioral advice and a checkup.

Illness may be to blame. Any condition that limits your cat's mobility, including blindness (see Chapter 4) or lameness (see Chapter 9), can make it difficult for her to reach the litter box for eliminations. Urinary incontinence (see Chapter 8) and anything causing diarrhea (see Chapter 10) can also make it hard for your kitty to stick to her litter box. In all cases of inappropriate elimination, you're smart to visit your vet for a routine checkup. He can diagnose and treat medical problems that trigger a lapse in bathroom habits.

Note: If your cat is an unfixed male and sprays urine on vertical household surfaces (like walls and drapes) while standing with upright tail, he is not urinating. He is spraying. Also known as marking, this is how a cat marks his territory. Although most common in unneutered males, unspayed females have also been known to mark territory. For intact animals, the best cure is fixing them. If the spraying still occurs in neutered pets, you may want to speak with your vet about the possibility of using hormonal drug therapy, such as progesterone. Another option is limiting the number of animals in a household, as the behavior comes from a cat's territorial marking instincts. Obviously, having only one cat would eliminate its need to mark territory.

Restlessness and Lethargy

Often there's a good reason your cat's activity level has gone up or down. In the case of lethargy, a record heat wave could be the cause: You'd be taking things slow, too, if you were wearing a fur coat. Restlessness can have many causes: Have you introduced a kitten into the household? Your grown kitty might be indulging in a second childhood now that she's got a playmate. Or, she could be physically hounding you (walking up to you repeatedly or standing up every time the new addition enters the room) because she needs reassurance that she won't be pushed aside.

If you can't, however, find anything in the cat's environment—no change in temperature, no new family members, no change in schedules—to explain the change in activity, a medical cause may be the underlying problem.

A restless cat is a fidgety cat. She may pace. She may sit, then stand, then lie down every few minutes. She may make repeated trips to the window. This may sound especially familiar if your cat hasn't been fixed. Should an uncastrated male detect a neighborhood female in heat, he will anxiously search for a way to get outdoors and find her. When in heat, a cooped-up female often paces, whines, and scratches at doors, trying to break free and meet a male. A cat who is restless, upset, or anxious could be hearing sounds of another animal outside in your yard, or perhaps a bird, squirrel, or raccoon in your attic. If your pet's restless behavior cannot be explained by any of these causes, your pet may be experiencing pain or discomfort from some internal medical problem—that is—nausea, intestinal cramps, muscle spasms, or pancreatitis.

Lethargy often looks like pure laziness, but it, too, can signal a condition. (Think of yourself: When you're not feeling well, moving around is one of the last things you want to do.) In fact, lethargy is a common symptom that accompanies a large number of conditions, including neurological conditions (see Chapter 3); cardiopulmonary ailments (see Chapter 7); digestive and reproductive conditions (see Chapter 8); and

19

bone and joint disorders (see Chapter 9). It can even tip you off to your cat's diminishing sight (see Chapter 4). Before you rush to the vet, you should always check for additional symptoms. Is there anything else your kitty is doing that's not normal for her? Coughing? Passing a discharge from one of her orifices? Shaking her head? Make a note of these symptoms and then take your cat to the vet: The more information he has, the more accurate his diagnosis will be.

CHAPTER 3

Head and Neck

The immune system and the nervous system make up two essential parts of your cat's body. The immune system, which stars the lymph nodes, works hard to protect the body from attack from foreign invaders—specifically, any illness-causing infectious organism. The problem with the immune system—from a cat owner's perspective—is that it works quietly. No high-drama symptoms here. So, if your kitty's lymph nodes are swollen, you won't know it unless you happen to run your hand over the area in a routine show of affection. This drives home the importance of taking a quick-yet-thorough daily inventory of your pet's health while playing with, petting, or grooming him.

The brain and spinal cord are known collectively as the nervous system. This network of neurons receives input from various body parts and from the outside world; it also transmits instructions throughout the body. These transmitted messages typically include communication between the nervous system and the outlying body parts regarding coordination, learning, memory, emotion, and thought. Though the signs of a problem with this system are a bit more obvious than they are with the immune system, they can still be overlooked. For instance, you may think that odd head-cock motion that your pet has recently made a habit of is just a cute kitten pose; in reality it may be a symptom called head tilt that indicates a neurological problem—a problem with the nervous system's functioning. On the other hand, some neurological signs that appear quite dramatic—seizures or fits, for instance—may actually be a passing symptom of unexplainable origin, which an owner will never see again.

Collapse

When the brain doesn't get enough oxygen, blood, or glucose (sugar: brain fuel), a cat can lose consciousness, a state that vets call **syncope**. The episodes are brief, usually lasting no more than 1 minute, and are akin to human fainting spells. In the rare case that a spell lasts longer than 3 minutes, however, the cat can die.

Collapse with Brief Loss of Consciousness and No Response to Voices or Touch

RELATED SYMPTOM: The cat appears to faint and remains unconscious for only 1 to 2 minutes.

POSSIBLE CAUSE: Has your cat suffered from head trauma? Has he been diagnosed with low blood pressure, a heart condition, or a metabolic disorder? A "yes" to any of these queries may point to **syncope,** a condition in which the cat loses consciousness due to a temporary lack of oxygen or glucose in the brain.

CARE: Keep the unconscious cat warm and the environment quiet until the animal wakes up. If he awakes in 3 minutes or less, make an appointment for the vet to examine your pet reasonably soon. To determine why your cat collapsed, your vet will perform a physical and neurological checkup and possibly a radiograph, an electrocardiogram, a blood test, and/or a brain scan. Conditions such as head trauma, low blood pressure, a heart condition, or a metabolic disorder can cause fainting. Treatment will depend on the underlying cause. Syncope is sometimes confused with mild seizures that produce unconsciousness without the more commonly associated violent jerking.

If 3 minutes have passed and your pet is still unconscious, carry him to the car and **take him to the nearest veterinary professional**—even if she's not your kitty's regular doctor.

PREVENTION: Address all head injuries and blood, heart, and metabolic disorders promptly.

Head Tilt and Lack of Coordination

Head tilt is a veterinary term describing a specific symptom that occurs when a cat tilts his head so that one ear is lower than the other. The cat's eyes may move continually in a jerky, side-to-side manner. Head tilt is a specialized symptom, usually indicating a condition involving the **vestibular mechanism** (which is made up of nerves of the inner ear, brain, and spinal cord and controls balance and posture) and/or the **cerebellum,** the region of the brain responsible for coordination.

The cat can also be affected by related "incoordination" symptoms: He may walk in circles (circling), lean on furniture or walls, stumble when walking, pick his feet up too high when walking (high-stepping), bob his head up and down when he eats, sway and/or fall sideways when standing, and appear mildly to severely disoriented. In fact, the presence of jerky, excessive movements often signals a disorder affecting the vestibular system or cerebellum. In some instances, incoordination is joined by facial paralysis, uneven pupil size, drooling, and difficulty eating.

Head-Bobbing, High-Stepping, and Swaying When Standing

RELATED SYMPTOMS: Your pet may also stagger and fall when getting up from a reclining position. Tremors may be present.

POSSIBLE CAUSE: Has your cat ever received a blow to the head? Has he been diagnosed with a viral infection or tumor? A "yes" to either of these questions may indicate **cerebellar disease,** a condition caused by any type of injury or inflammatory process involving the cerebellum, the part of the brain responsible for coordination of movement.

CARE: Take your pet to the vet, who will perform a complete physical and neurological exam to reach a diagnosis, including blood tests and possibly a skull radiograph or brain scan. If your kitty does have a cerebellar disorder, your vet will treat its underlying cause. For example, an infection warrants antibiotics, whereas a tumor can, in some instances, be removed. Anti-inflammatory drugs may be needed to reduce inflammation and pressure on the brain. To keep your cat safe from everyday dangers (such as falling down stairs), confine him to a safe, comfortable area of the house.

PREVENTION: There is no prevention.

Head Tilt and Jerky Side-to-Side Eye Movements

RELATED SYMPTOMS: In addition to tilting his head, your cat may also veer into furniture or walls, stumble when walking, sway sideways when standing, fall for no apparent reason, high-step, and/or circle. His eyes may move rapidly (in a manner some people might call "typewriter-like"), and he may seem disoriented and weak. The symptoms usually come on suddenly and may be accompanied by nausea, vomiting, facial paralysis, and drooling.

POSSIBLE CAUSE: Has your cat ever suffered an injury to the head that could have caused brain trauma? Has he been diagnosed with a brain tumor or ear infection? Is he an elderly cat? A "yes" to any of these questions may indicate a **dysfunction of the vestibular system,** a network of nerves in the ear, brain, and spinal cord that govern balance and orientation.

CARE: In mature animals, a vestibular condition will often disappear without treatment within 1 to 6 weeks. Still, it doesn't hurt to have your pet checked by a veterinary professional, who will use radiography and/ or a brain scan to rule out a brain tumor or infection. The exact care your vet gives depends on what is causing the vestibular condition, although sedatives and indoor confinement may be prescribed to keep the cat calm and safe.

PREVENTION: Addressing all head injuries and ear infections promptly can limit your pet's chances of developing a vestibular condition.

Staggering Gait, Moodiness, Aggressive Behavior, and Excessive Salivation or Frothing

RELATED SYMPTOMS: You may notice changes in your cat's personality. The cat may swat, lunge, or bite without provocation, lose his appetite, stop drinking water, howl frequently, destroy property, act restless, and/or be overly timid. The cat may have an absent gaze, and his pupils may be unevenly dilated. Other physical signs you may notice include a drooping lower jaw, excessive drooling, seizures, and wandering.

POSSIBLE CAUSE: Has your pet recently spent an unsupervised amount of time outdoors? Is there a wound on his body that could be the result of a bite? If you answered "yes" to both and he had not already been vaccinated for the disease, he could have contracted **rabies.** The rabies virus, which lives in saliva, can be passed on to your pet if he is bitten by an infected skunk, raccoon, neighborhood kitty or pooch, or wild animal. The virus works by attacking the nerve tissue, hence the disease's many neurological symptoms. The incubation period can be as long as 6 months or as short as 2 hours. Death usually occurs 3 to 10 days after the onset of symptoms.

CARE: Not only does rabies pose a threat to humans, but a loose, rabid cat can infect several other neighborhood animals, sparking a mini-epidemic. If you suspect your pet has been infected with the virus, try to lure him to a secluded room and—this is going to be heartbreaking—have as little contact with him as possible until you've called your vet and the animal-control officer. The local authorities will pick up your cat for quarantine and observation. If, after being watched, your cat appears to have rabies, he will die quickly or be put to sleep. Unfortunately, the only way to know for sure if your pet had rabies is to perform an autopsy of the animal's brain and salivary glands.

PREVENTION: Make sure your kitty is vaccinated against rabies, and supervise his outdoor play.

Wobbliness When Standing or Walking and Incoordination of Limbs

RELATED SYMPTOMS: Your cat's head and neck movements may also appear uncoordinated, almost as if one is moving with no consideration of what the other is doing.

POSSIBLE CAUSE: Has your cat recently been diagnosed with any of the following: anemia, a heart disorder, a respiratory illness, or a disc or cervical-cord disorder? Is your pet taking tranquilizers, antihistamines, or anticonvulsants? A "yes" to any of these questions may point toward **ataxia,** a sign that reflects a problem in your cat's nervous system. Ataxia is actually a symptom, not a disease, which results from hampered communication between your cat's brain and body, making it difficult for the body to carry out any orders the brain makes for movement. The symptom is brought about by the damage of specific neurological illnesses or by a reaction to certain medications. Inner-and middle-ear disease can also produce ataxia.

CARE: Take your pet to the vet, who will diagnose the condition after performing a series of tests, including a neurological exam, a blood test, and a radiograph. If a specific illness is behind your kitty's ataxia—for example, a degenerative spinal-cord disease—that disorder will be addressed first. Generally, once that primary illness is treated, ataxia disappears. If your kitty must regularly take a prescription medication that may happen to be causing the ataxia, your vet will explore other medication options.

PREVENTION: There is no known prevention.

Paralysis

Because cats are active creatures, finding your fabulous feline suddenly unable to move a body part may come as a scary surprise. Vets refer to this condition as **paralysis,** and it can affect the head, the neck, one or more legs, the tail, the spine, or a combination of body parts. Paralysis results from nerve-cell impairment, which may be caused by trauma, infection, a broken bone, a herniated disc, a brain or nervous-system disorder, a tumor—or even a blood clot or an inherited ailment. Should you notice that your kitty is unable to move a body part, look for other possible signs that something is wrong, such as pain, changes in eating or elimination habits, or any swollen areas. Make a note of these symptoms and visit your vet, who will hopefully be able to make a diagnosis using the information you've provided.

Muscles That Appear Frozen, Stiff-Legged Gait, and Lying on His Side

RELATED SYMPTOMS: The cat's ears may be frozen in an erect stance, his mouth may be drawn back, and his eyes may be narrowed. He may have difficulty swallowing and may drool. You may see vertical lines on his forehead. The cat may also have a fever.

POSSIBLE CAUSE: Did your cat receive some type of puncture wound in the last 48 hours? If so, he may have **tetanus,** although you should note that tetanus is uncommon in cats. The condition is caused by bacteria that enter the body through a puncture wound. These bacteria release a nerve toxin that causes the body's muscles to stiffen.

CARE: Take your cat immediately to the vet, who will quickly administer tetanus antiserum and antibiotics. To help further relax frozen muscles, strong sedatives are given. Be aware, however, that tetanus can be fatal to cats, despite veterinary care.

PREVENTION: Supervise your cat's play. If your cat develops a deep wound, thoroughly clean and flush the area as soon as possible. Felines are highly resistant to the bacterium that causes tetanus (much more so than humans are) so they rarely develop the condition and are rarely vaccinated for it. But if you live in an area with livestock, talk to your vet about a tetanus shot: The tetanus-causing bacterium thrives in manure and manure-contaminated soil.

Paralysis of the Face

RELATED SYMPTOMS: Your cat's eyes, eyelids, ears, nostrils, and/or lips are stiff and unmoving. One or both sides of the face will be affected, and you may notice asymmetrical drooping of the ears or lips and an inability to blink on the affected side(s). Your kitty may be messier than usual while eating, and also drool.

POSSIBLE CAUSE: Has your cat recently been diagnosed with an ear condition? Has he been involved in an accident that resulted in some type of facial trauma? A "yes" to either question may indicate **facial nerve paralysis,** an ailment in which the facial nerve becomes impaired, making facial movement difficult. Middle-ear infections are the most common cause of facial paralysis in cats.

CARE: Take your kitty to the vet, who will diagnose the condition after performing a radiograph. Whether the condition can be cured depends on its cause. For example, facial paralysis due to a middle-ear infection is treatable. To help your cat's immune system deal with the cause of his paralysis, give him vitamin C (calcium ascorbate), beta-carotene, and vitamin E (see Appendix E: List of Recommended Dosages, pp. 169–189).

PREVENTION: Promptly address ear infections and any facial trauma. Also, supervise outdoor play to keep your cat from becoming involved in an accident.

Seizures and Tremors

Seizures, fits, convulsions—call these episodes of uncontrolled movement what you like, they are still frightening—just ask anyone who has ever seen an animal or human in the throes of such a fit. In a generalized seizure, the victim stumbles to the ground, limbs stiffen and twitch, eyes roll back, the head shakes, the neck arches back, the jaws mash together, and drooling and loss of bladder and bowel control may occur. A local attack (called a partial seizure) is limited to spasms of a single body part or region, such as the facial muscles, head and neck, hindquarters, or just one leg. Fortunately, an individual incident is usually brief, lasting from 3 seconds to 3 minutes or, in rare instances, up to 1 hour. A cat may have only one episode during his lifetime, or he may have several distinctly separate spastic episodes during 3 minutes or 3 hours, 1 time a week, or 1 time every month. Occasional mild, atypical seizures may show no muscle twitching and may appear as momentary fainting spells.

Should your pet have such a fit, quickly create a calming environment: Clear away all sharp objects, try to surround the animal with a cushion of soft bedding, remove any constrictive collars, turn off a loud stereo or TV, and keep your hand away from the victim's mouth (or you may get your fingers caught between teeth). No, your pet won't swallow his tongue.

Animals rarely die during a seizure, although fits are considered life-threatening—and in need of veterinary attention—especially when a number of separate seizures are quickly repeated without the cat gaining consciousness between each one. Also, if a single fit lasts longer than 3 minutes, medical help should be sought. Wrap your kitty in a cocoon of heavy blankets and **go straight to the nearest vet clinic or hospital**.

As dramatic as this spastic display is, whatever prompted the seizure should concern you most. Of course, one major cause is **epilepsy.** Other conditions connected with more generalized seizures include **liver** and **heart disease, metabolic system abnormalities, brain tumors** and **trauma, infectious diseases,** and **poisoning.** When the seizure has passed—or during the episode if the convulsions are prolonged or repeat themselves—take your cat to the vet. She can control the seizure and then attempt to discover what underlying condition has brought on the fit.

Unlike seizures, a **tremor** is an involuntary movement of either the entire body or a part of the body. The action is constant, rhythmic (almost in a to-and-fro fashion), and happens while your cat is conscious. A seizure, on the other hand, can occur while the cat is unconscious.

Depending on the accompanying symptoms (such as apathy or a change in posture), tremors usually indicate muscle weakness or a brain disorder. Note any of these additional signs and report them to your vet.

Spasms and Uncontrolled Movement of the Head, Eyes, and Mouth

RELATED SYMPTOMS: Muscles of the facial area will twitch. The spasms may or may not spread to include the entire body.

POSSIBLE CAUSE: Has your cat been diagnosed with a brain lesion? A **partial seizure** is common in animals with brain lesions.

CARE: During the fit, remove all sharp objects from the area and make your cat as comfortable as possible (see Seizures and Tremors, pp. 27–28). Once the seizure has passed—or during the fit if it lasts more than 3 minutes—take your pet to the vet, who will perform a neurological examination, take a radiograph, and/or perform a brain scan to determine what is causing the seizure. If a brain lesion is what prompted the partial seizure, a veterinary neurologist will have to determine whether it is operable or not. If not, she will attempt to control the seizures with medication.

PREVENTION: There is no known prevention.

Swollen Lymph Nodes

When feeling achy and sick, how many times have you reached up to your throat to check the size of the glands tucked inside? What you're feeling for are your **lymph nodes.** These round organs range from the size of a pea to the size of an olive, and they are nestled under the skin at various sites throughout the body, including under the arms, behind the knees, and in the groin. These glands produce **lymphocytes,** illness-fighting white blood cells that produce antibodies and help protect the body from invasion by bacteria or other organisms. When busy churning out these warrior blood cells, lymph nodes often swell and become tender to the touch.

All this goes for your kitty, too. Enlarged glands—which you will probably notice while petting your cat—can mean your cat's body is hard at work trying to fight off an illness. However, to find out which illness, you should check for the presence of any other symptoms and report them to your veterinarian.

Enlarged Lymph Nodes, Pale Mucous Membranes, Recurring Infections, Apathy, Loss of Appetite, and Fever

RELATED SYMPTOMS: Your cat has no appetite and vomits occasionally. She may also be constipated and/or have diarrhea, and she may be prone to gingivitis, skin lesions, and/or upper-respiratory conditions. Symptoms generally last from 2 to 16 weeks.

POSSIBLE CAUSE: Is your cat unvaccinated against the feline leukemia virus and has she had access to other cats—possibly feral felines—at any time within the last 5 years? Did her mother test positive for the feline leukemia virus? There's a chance she has picked up the **feline leukemia**

virus. FeLV, as the illness is known in veterinary circles, belongs to the retrovirus family, as does the human AIDS virus. This kinship explains FeLV's insidious modus operandi: Cloaking itself in the body secretions of an infected animal, the virus then enters a new animal's body through any number of ports: the eyes, ears, mouth, nose, sexual organs, or a wound. (In other words, an uninfected cat can get FeLV just by hanging around an infected friend, sharing her water bowl, taking part in mutual grooming, being sneezed on, and so on.) Once inside the new host, the virus gets busy breaking down the animal's immune system.

Infected cats succumb to a wide range of cancers and secondary infections, including feline infectious peritonitis, feline viral respiratory-disease complex, chronic cystitis, periodontal disease, and various bacterial infections.

To make matters worse, FeLV can have a long incubation period. It can incorporate itself into a cat's genetic material, remaining dormant—meaning it doesn't cause disease symptoms—for long periods of time. As a result, an infected cat may not show any signs of illness for years—though most begin to show signs within 3 weeks up to 5 years of exposure.

CARE: There is no cure for FeLV. If you suspect your cat may be infected, take her to the vet, anyway. Testing saliva, tears, and/or blood will uncover the virus's presence. Should your pet test positive, your vet will concentrate on treating whatever clinical signs, cancerous conditions, and/or secondary infections crop up. Unfortunately, there is no way to tell how long an infected cat will live. She may spend half a decade suffering from the mildest of symptoms, or be leveled by a fatal cancer just months after she's been diagnosed with FeLV.

If yours is a multiple-feline home, the affected cat's housemates must also be brought in. There is a good chance that they, too, are carrying the disease. To build up your cat's immune system, give her antioxidants, including the vitamins A, C, and E and beta-carotene. Dimethylglycine (available at health-food stores) is helpful, too. You also may want to give your kitty fresh aloe vera juice containing acemanin, an antiviral agent (see Appendix E: List of Recommended Dosages, pp. 169–189).

PREVENTION: A feline leukemia virus vaccine is available: Make sure your cat receives it if he goes outdoors. More commonsense advice: Do not add a second (or third, or fourth) cat to a feline household without first obtaining a FeLV-free diagnosis from your vet. Last, don't allow a

FeLV-positive male or female to breed: Their kittens may be born with the disease.

Enlarged Lymph Nodes, Fever, and Lethargy

RELATED SYMPTOMS: The enlarged lymph nodes are most often found in the chest and abdomen, but may also be located under the lower jaw, in front of the shoulder blades, behind the knees, and in the groin. Your cat may show no interest in food and lose weight. He may drink more than usual. His abdomen may appear swollen and may be tender. He may vomit and/or have bouts of recurring diarrhea. He may also cough and have difficulty swallowing. The eyes may appear cloudy and skin may be scaly.

POSSIBLE CAUSE: Is your cat older than three years? Has he been diagnosed with the feline leukemia virus (FeLV)? (See the preceding section, Enlarged Lymph Nodes, Pale Mucous Membranes, Recurring Infections, Apathy, Loss of Appetite, and Fever, pp. 29–31.) Your pet may have **lymphosarcoma,** also known as **cancer of the lymph nodes.**

Although lymphosarcoma also appears in cats who do not have the feline leukemia virus, it is more common in animals who have been diagnosed with FeLV. In other words, lymphosarcoma is not caused by FeLV: The virus simply makes cats more prone to developing it. A good analogy is pneumonia in humans with AIDS: AIDS doesn't cause pneumonia, but AIDS patients are more prone to developing pneumonia than those without the disease.

Now that you know who is susceptible to lymphosarcoma, you now need to become acquainted with the condition itself. Lymphosarcoma is a type of cancer that metastasizes, which means it spreads to other areas of the body, such as the chest, gastrointestinal tract, skin, eyes, and central nervous system. Your cat's symptoms will greatly depend on which lymph nodes the cancer is in and to what part of the body—if any—the cancer has spread.

CARE: Take your pet to the vet, who will take blood samples and perform a tissue biopsy to determine whether lymphosarcoma is present. If it is, and if the cancer is limited to specific external lymph nodes, those nodes will be removed. If, however, the malignancies have spread, chemotherapy and/or radiation treatments will be needed to help slow down the cancer.

In cats, lymphosarcoma is considered potentially "curable" with chemotherapy. (Though the number of vets who conduct chemotherapy in their offices is increasing, taking your kitty to a nearby veterinary hospital or university veterinary clinic for regular treatments may be necessary.) Bear in mind that chemotherapy in animals is not as dramatic—or stressful—as in humans. If your kitty goes into—and stays in—remission for 8 months, the chances are good that he will go on to live 4 or more cancer-free years. At home, you can keep your cat's immune system strong by providing him with vitamins A, C, E, dimethylglycine, and beta-carotene. Fresh aloe vera juice containing the antiviral agent acemanin is also recommended (see Appendix E: List of Recommended Dosages, pp. 169–189).

PREVENTION: Although lymphosarcoma cannot be prevented, having your kitty tested for and vaccinated against the feline leukemia virus will lower the chances of your cat developing cancer. Should your pet develop lymphosarcoma, his life may be saved if the cancer is caught and treated in its early stages.

Swollen Lymph Nodes, Fever, and Neurological Abnormalities

RELATED SYMPTOMS: Your cat may stagger when walking, have a stiff gait, appear dazed, and/or tilt his head. The cat's eyes may be inflamed. He may also chew or nip at his skin. You may actually see dark purple-gray skin tags or mahogany-colored bugs, white dots, and/or black flecks clinging to individual hairs. The cat's skin may have dry patches or pimplelike sores.

POSSIBLE CAUSE: Does your pet spend time in wooded or rural areas? Have you recently found a tick on your kitty? There's a chance he may have contracted **Lyme disease.** Although rarely found in cats, this illness can be caused by bacteria and is transmitted to animals by ticks.

CARE: If you suspect your cat has Lyme disease, take him to the vet, who will perform blood tests to discover the condition's presence. However, you should note that Lyme disease in cats is both rare and difficult to diagnose. Antibiotics are generally administered for 2 to 4 weeks to treat the disease, but it is not known if they completely eradicate it. For this reason, their success rate is difficult to measure. If you are feeling under-the-weather, schedule a checkup for yourself as well. Although you can't get Lyme disease from your pet, you can get it from the local ticks who infected your kitty. To strengthen your pet's immune system, provide him with a high-quality diet, supplemented with antioxidants such as beta-carotene and vitamins A, C, and E (see Appendix E: List of Recommended Dosages, pp. 169–189).

PREVENTION: Unfortunately, there is no Lyme disease vaccine for cats. Keep your cat away from wooded areas. Because infection does not occur until the tick has been attached and feeding off your cat's blood for some time, removing all ticks immediately is important (for tick-removing instructions, see Chapter 6, p. 65) and dust the cat's fur with tick powder. You can also apply monthly flea and tick drops such as Frontline to ward off these parasites.

Enlarged Lymph Nodes, Unresponsive Chronic Diarrhea, Weight Loss, Appetite Loss, Bad Breath, and Chronic Gingivitis

RELATED SYMPTOMS: Your cat may act disoriented or be either more shy or more aggressive than usual. You may notice she is prone to developing recurring infections of the skin, respiratory tract, bladder, and other body parts.

POSSIBLE CAUSE: Does your cat have access to other cats? Have you noticed a deep bite wound on your cat within the last 5 years? She could have the **feline immunodeficiency virus,** also known as **feline AIDS** and **FIV.** Similar to both the feline leukemia virus (see the section, Enlarged Lymph Nodes, Pale Mucous Membranes, Recurring Infections, Apathy, Loss of Appetite, and Fever, pp. 29–30) and human AIDS, FIV attacks your cat's immune system, leaving her susceptible to a wide range of infections and cancers.

Unlike FeLV, feline immunodeficiency virus is rarely spread via casual contact with body fluids such as saliva and urine. And unlike human AIDS, the feline version is not transmitted via sexual encounters. It is spread when an infected cat bits deep into an uninfected kitty's tissue, injecting infected saliva directly into the bloodstream.

Just like FeLV, FIV can have a long incubation period. It can incorporate itself into a cat's genetic material, remaining dormant—meaning it doesn't cause disease symptoms—for long periods of time. As a result, an infected cat may not show any signs of illnesses for years, though most begin to show signs within 3 weeks up to 5 years of exposure.

CARE: There is no cure for FIV. That said, you still should take your cat to the vet if you suspect the condition. A blood test can expose the virus's presence. Incidentally, most vets also perform a test for FeLV at the same time; the two illnesses often occur simultaneously in cats.

Should your pet test positive, your vet will concentrate on treating whatever clinical signs, cancerous conditions, and/or secondary infections crop up. Unfortunately, there is no way to tell how long an infected cat will live. She may spend half a decade affected by the mildest of symptoms, or be leveled by a fatal cancer just months after she's been diagnosed with FIV.

If yours is a multiple-feline home, the affected cat's housemates also must be brought in for testing. (For those who are wondering—or worrying—FIV is not zoonotic, meaning that it cannot leap over the boundaries between species to infect humans.) To keep your kitty's immune system strong, give him vitamins A, C, E, dimethylglycine, and beta-carotene. Fresh aloe vera juice containing acemanin (an antiviral agent) is also recommended (see Appendix E: List of Recommended Dosages, pp. 169–189).

PREVENTION: There is no vaccine for FIV, thus the best—indeed, the only—prevention is to keep your cat from being bit by an FIV-positive feline. This is easiest done by keeping your cat indoors, or allowing her outdoors only when you can watch her activities.

Do not add a second (or third, or fourth) cat to a feline household without first obtaining a FIV-free diagnosis from your vet.

CHAPTER 4

Eyes, Ears, and Nose

As clichéd as it sounds, the sensory organs really are your pet's windows to the world around her. Given the importance of eyes, ears, and nose in your kitty's life, she'll probably let you know there's a problem as soon as she knows herself. For instance, if she gets a foxtail stuck in her ear, she'll give you clues to decipher—in this case scratching her ear and shaking her head in attempt to dislodge the irritant. Many symptoms may even be visually obvious to you, such as a discharge from the nose, waxy buildup in the ear, or a growth in one of the eyes.

Although conditions involving the sensory organs often come on suddenly, some develop gradually. Ailments involving the eye, such as cataracts and glaucoma, can be especially slow in evolving. Just like a human who is surprised during a routine eye exam to learn he has a worsening vision impairment, by the time your kitty's sight is severely hampered, she will have had enough time to develop ways to adapt to the situation. It is you, her owner, who will think it odd that your once-athletic cat now ignores half the balls you throw to her or that she doesn't enjoy chasing her favorite pull-toy the way she once did.

Abnormally Shaped Eyes

A misshapen eyeball is a dramatic signal that something is wrong with the eye. Perhaps the most obvious abnormality is the bulging eye. Though most notably linked with glaucoma, a bulging eye can also be a defect present at birth, a side effect of severe dental disease, or the result of a tumor growing in the eye socket behind the eyeball. In rare cases, an eye may wither and become smaller, as in cases of severe inflammation due to **endophthalmitis.**

In most cases, an abnormally shaped eye is an uncomfortable eye. Your cat may blink more than normal, have watery eyes, and in some instances experience vision loss. If an eye bulges or enlarges to an extreme, the eyelids may not be able to close and the eye will tend to dry out.

Enlarged Eyeball That Bulges from the Eye Socket, with a Dilated Pupil and Tearing

RELATED SYMPTOMS: The retina may have a faint green tinge. You may notice inflammation of the entire eye. Vision may seem diminished, and the eye is often very painful.

POSSIBLE CAUSES: Has your cat ever been diagnosed with an eye condition, such as lens displacement, retinal atrophy, or cataracts? Has she ever experienced an eye infection or injury? A "yes" to either of these questions may indicate **glaucoma**, a condition caused by increased fluid buildup and pressure within the eyeball, which, in turn, damages the optic nerve, possibly resulting in blindness. In cats, glaucoma is typically a secondary condition, meaning that glaucoma often develops after another ailment—such as those just listed—is present.

CARE: Take your pet to the vet, who may suspect the condition after a physical exam. He will then measure the pressure of the eyeball with a tonometer. Drugs can relieve the fluid buildup, and thus the pressure and pain, in the eye. Several surgical techniques, including cryosurgery (use of subfreezing temperature), have been developed to relieve the intraocular pressure. Your vet may suggest removing the entire eye if it has been severely damaged by glaucoma.

PREVENTION: Address all eye conditions promptly. A red, inflamed eye may be the earliest indication that glaucoma is present. The earlier the diagnosis, the better the chance to prevent the loss of your cat's vision and possibly the loss of her eye.

Eye Pops from the Socket

RELATED SYMPTOM: No other symptoms. Note that the eye doesn't actually fall out of the head because it is held in place by the lid.

POSSIBLE CAUSE: Is your cat a short-nosed breed, such as a Persian or Burmese? Were you recently cuddling your pet or applying any type of pressure to her head? She may be suffering from a **proptosed eye**, also referred to as **proptosis of the eye**, a condition where the eye literally pops out of the socket.

CARE: As unnerving as this experience is likely to be, calmly take your pet to the vet—moving her as little as possible. If the eyeball has very recently popped from the socket, your vet can push it back in place by hand. But if some time has elapsed between the eye's escape from the socket and your vet visit (maybe you didn't discover the injury right away), there may be some fluid buildup in the socket. Your vet will drain this, then put the eye back into place.

PREVENTION: Use care when holding your cat's head still for grooming or giving medication.

Cloudy Appearance to the Front of the Eye

Eye conditions are quite common in cats, especially older animals. In fact, there's a good chance that you've seen a kitty with an eye that appeared less than bright and clear: Perhaps a milky haze covered the entire cornea, an opaque, off-white curtain obscured the **retina**, or spots of white dotted the cornea.

This cloudy appearance to the normally transparent eye, called an **opacity** by vets, primarily affects two parts of the eye: the **cornea** (the transparent structure that covers the front of the eyeball) and the **lens** (a transparent biconvex sphere located behind the cornea and pupil that focuses light rays upon the retina). These irregularities can be located either on the outside of the eye or under the surface. They can be white, gray-white, green-white, blue-white, yellow-white, or crystalline in appearance. They may be slightly see-through or completely opaque, and they can cover either the whole cornea or just bits of it.

So what are these opacities—and what are they doing on your cat's eye? They may actually be: scars, in the case of **eye injuries**; inherited problems, such as **corneal dystrophy**; infection-related conditions, causing an inflammation of the cornea; or age-related problems, as in some instances of **cataracts**. A vet needs to look at the eye to reach a diagnosis and devise a treatment plan; however, most of these conditions develop gradually and rarely qualify as extreme emergencies.

Cloudiness or Milkiness of the Normally Transparent Cornea (Corneal Opacity) or a Blackish Film Covering the Cornea

RELATED SYMPTOMS: One or both eyes may be affected and the cloudiness may also be accompanied by crystalline grayish, whitish, or silverish spots. You may notice tearing and redness of the eye. Furthermore, your cat may be very sensitive to bright lights and might blink excessively and/or rub or scratch at her eye.

POSSIBLE CAUSES: Has your cat been suffering from an eye condition caused by a bacteria or virus? Has she ever experienced an eye injury? Does she ever fight with other cats? Is she a short-nosed breed or breed mix, such as a Persian or Burmese? Did one of her parents have corneal problems? A "yes" to any of these can point to any number of **disorders of the cornea**, including acute inflammation (**keratitis**), **corneal erosions** or **ulcers, dystrophy,** or **degeneration**.

CARE: Some corneal conditions are not sight-threatening, whereas others can cause blindness if left untreated. Therefore, it's a good idea to see your vet quickly. Corneal erosions and ulcers are treated first with antibiotics and then with anti-inflammatory drops. If the ulcer resists healing, surgery may be necessary to remove dead tissue. Thicker, more superficial opacities can be surgically peeled off the front of the cornea. Corneal ulcers that heal may leave a permanently scarred area over a portion of the cornea. To encourage healing, give your cat vitamin A and vitamin C (calcium ascorbate) (see Appendix E: List of Recommended Dosages, pp. 169–189). Placing a drop of cod liver oil directly in the eye 1 to 2 times a day is also recommended.

PREVENTION: Address all eye conditions promptly. To prevent your cat's cornea from being scratched, keep your pet away from unknown—and unfriendly—cats, high grass, sharp objects, high winds, and dusty conditions.

White to White-Blue Opacity of the Lens, Accompanied by a Dilated Pupil

RELATED SYMPTOMS: The lens opaqueness involves only the pupil. (The pupil, or pupillary area, is the hole or empty space created by the iris muscle, the colored portion of the eye, as it opens and closes to let more light pass through the lens. The lens then takes the light that has passed through the pupil and focuses it on the retina. Therefore, lens opacities usually appear deeper than corneal opacities.) The eye is usually not painful and there is little to no tearing or redness.

POSSIBLE CAUSE: Is yours an older cat? Did one of her parents have cataracts? Has she been diagnosed with diabetes? Has she ever been diagnosed with retinal atrophy or a displaced lens? A "yes" to any of these questions may indicate that your cat has a **cataract**, an ailment in which the lens of the eye becomes opaque, often causing blindness. Although uncommon in felines, it can occur.

CARE: Have your vet give the cat a physical and ophthalmic exam in order to diagnose the problem. If the cat is elderly and the cataract affects only one eye, your vet may choose to leave the situation alone. However, because a cataract can grow to cover an increasingly large portion of the eye, your vet may want to remove it. Surgical removal of the cataract is now done with a hi-tech device that sucks the opaque lens material out from the lens capsule. Cats can see adequately without a lens, therefore there is no reason to have a prosthetic one put in (as is sometimes done in humans).

PREVENTION: Treat all metabolic diseases and eye conditions promptly. To help slow the progression of cataract formation, provide your cat with a high-quality, chemical-free diet, supplemented with an amino-acid preparation (your vet can recommend a single, multiamino acid supplement) containing histidine, phenylalanine, and taurine. If you feed your cat an exclusively vegetarian diet, you should know that it is absolutely essential to supplement it with taurine. You may also want to give your kitty vitamin E and vitamin C (calcium ascorbate) (see Appendix E: List of Recommended Dosages, pp. 169–189)

Diminishing Sight

There are many types of blindness: partial, full, night blindness, blindness in daylight caused by sun sensitivity (photophobia), and blindness in one eye. Blindness can also have many causes: heredity, degeneration, infection, poor diet, or trauma. Diminished sight can also be a secondary result of a primary eye disorder such as cataracts or glaucoma. Age is also a factor: As most adult humans know, sight becomes less sharp with time; the same is true for cats. Both cataracts and sclerosing (hardening) of the lens (which is often confused with cataracts) are age-related problems that result in diminished sight.

If you notice your kitty bumping into things, not seeing movements in her peripheral area of vision, exhibiting difficulty with navigating in darkness, missing balls during a game of chase-the-ball, and hesitating to explore new or dark environments, her sight is probably failing. Visit your vet so that he can determine what problem is behind your cat's impaired vision and whether the problem can be reversed—the latter depends greatly on the former. For instance, glaucoma-reduced vision typically improves once the glaucoma has been treated, whereas impairment caused by a retinal condition is often irreversible. If cataracts are removed, the cat's sight will improve, but it will still be far from perfect.

Blindness in Dim Light, Dilated Pupils, and a Marked Preference for Well-Lit Areas

RELATED SYMPTOMS: You may hear your cat bumping into things at night. You may also notice her reluctance to go out into the yard at night and a tendency to huddle near your legs in low light. In advanced cases, daytime vision might also be affected and you might notice your cat missing a ball when playing chase-the-ball.

POSSIBLE CAUSES: Has your cat been diagnosed with feline toxoplasmosis, feline infectious peritonitis, lymphoma, cryptococcosis, or a systemic fungus infection? Does she suffer from hypertension? Has she ever had an eye injury? Has she been on a lower-quality food that may be deficient in the essential amino acid, taurine? Is she a Persian or Abyssinian? She may have one of several types of **retinal diseases**.

Retinal diseases make it difficult for the eye to interpret whatever light it receives, which is why these diseases first affect night vision, then progress to the point where daytime vision becomes blurred. Among the retinal diseases, three are most common in cats. **Progressive retinal atrophy**, which eventually leads to total blindness, is often hereditary in Persians and Abyssinians, though it can strike an animal of any breed for seemingly unknown reasons. **Retinitis** is usually provoked by either an illness (such as toxoplasmosis, cryptococcosis, or hypertension) or an eye injury that causes the retina to become inflamed. When inflamed, the retina's light receptors begin to degenerate, first affecting night vision, then day vision. **Retinal degeneration** is usually caused by a taurine deficiency. When the cat's body doesn't get enough of this essential amino acid, the central area of the retina starts to degenerate, leaving peripheral vision but making it hard for your cat to see stationary objects.

CARE: Take your pet to the vet, who will examine the eyes and retina using a special instrument called an indirect ophthalmoscope. He may also refer you to a veterinary ophthalmologist to perform an electroretinogram to diagnose a retinal condition.

Whether your cat can be cured depends on the type of retinal disease with which she has been diagnosed. Retinal atrophy progressively advances into total blindness. (As disconcerting as it is to think your cat may go blind, many cats live happy, nonsighted lives when limited to familiar surroundings.) Retinitis, however, may be halted, controlled, or even slightly corrected, if the illness or situation that originally prompted the eye condition is treated. Retinal degeneration, too, can be halted and possibly corrected by introducing taurine into the kitty's diet.

PREVENTION: Although you can't prevent every type of retinal condition by giving your cat taurine, you can neuter your pet to ensure that other animals with a hereditary retinal condition aren't brought into the world.

Eye Area Growths

Bumps or lumps on your cat's eyelid or lashline are usually **sties, pimples, warts,** or **tumors.** They may or may not ooze a liquid, and most are slow in developing. Moreover, they rarely affect sight, but depending on how close they are to the cornea itself, they may irritate the eye and cause redness. These growths rarely constitute an emergency, but you'd be wise to have a vet examine your cat's eye and, if necessary, remove the mass. In the meantime, keep the eye clean and free of debris by flushing it with a mild boric-acid solution or with a product such as Visine or Murine.

One or More Red, Black, or Pink Lumps Along the Lashline or Eyelid

RELATED SYMPTOMS: One or both eyes may be affected and hair loss may occur around the lumps. The eye is usually not painful.

POSSIBLE CAUSES: The preceding symptoms describe **sties, warts, chalazions, cysts,** and **tumors of the eyelid.** They are not related to or caused by any other diseases, and although they are found more often in older animals, cats of any age can develop such lumps.

CARE: Take your pet to the vet for a diagnosis. Care depends on what

41

type of growth or growths your cat has. Sties can be treated with a regimen of hot compresses and antibiotic ointment. Warts and chalazions are usually surgically excised. In the case of tumors, a biopsy will be performed after the growth has been removed to determine whether the growth is cancerous—*most lid tumors are, in fact, benign.* If the tumor does happen to be malignant, your vet will probably limit treatment to surgical removal. Since cancerous growths in the eye area tend to spread slowly and invade only local tissue, chemotherapy is usually unnecessary.

PREVENTION: There is no way to prevent these growths. However, early detection when the lumps are very small makes for a more cosmetically pleasing surgical excision.

Eye Tearing

Eyes water when they are irritated. The cause of the irritation can be anything from a speck of dirt, to a sty, to an underlying medical condition. If the tearing is in both eyes and is the only noticeable symptom, the problem is probably not an emergency. Think back to what your cat was doing before the tearing began. Was she resting in a part of the house that was being cleaned? Was she outdoors during windy weather? If so, there's a good chance that the eye is trying to flush out a piece of grit. If, however, the tearing comes on slowly, is in one eye only, and/or is accompanied by any change in the appearance of the eye or a growth in the eye area, the problem may be more serious than a lodged speck of dust. Go to the vet as soon as possible: Eyes are delicate organs, and eye problems can go from bad to worse very quickly.

When a medical condition is causing your kitty's eyes to water, "companion symptoms" are often present. Redness is a prime example. A red, inflamed eye can be the sign of a more serious problem. Pain as evidenced by squinting or blinking should receive immediate attention. Carefully note any of these symptoms: They are what a vet uses to help distinguish one eye condition from another, and thus reach a diagnosis.

Last, if you have a light-haired Persian, you may notice that everyday tearing—for example, the kind prompted by a piece of dust in the eye—seems to stain the fur by the inner corner of the eye directly under the lower lid. In most cases, this is completely normal and does not indicate an eye ailment. If, however, you're bothered by how this staining looks, talk to your veterinarian. Low doses of certain oral antibiotics (in liquid form) seem to clear up the staining in some animals, though no one knows why.

Excessive Tearing with Sticky, Yellowish Discharge in the Inner Corners of the Eyes, Accompanied by a Number of Eyelashes That Appear to be Growing into the Eyes

RELATED SYMPTOMS: Both eyes are involved.

POSSIBLE CAUSE: Do you notice lashes growing toward the eyeball? Is your cat a flat-faced breed, such as a Persian or Himalayan? Your pet may have **distichiasis**, a somewhat uncommon condition in which lashes, instead of growing away from the eye, actually grow into the eye. This results in irritation, inflammation, and infection.

CARE: Take your pet to the vet, who can diagnose the condition after a physical examination. The stray lashes will be removed using cryosurgery (which uses subfreezing temperature to kill the hair follicles). Antibiotics will be given to clear up any existing infections. (Caution: *Do not* attempt to trim these hard-to-see lashes at home! Leave lash removal to a veterinary professional.)

PREVENTION: There is no known prevention. However, to keep problems at bay, give your vet a call should you notice a lash—or lashes—growing in the wrong direction. He may choose to remove the offending hairs before they get long enough to touch the eyeball.

Excessive Tearing with Sticky, Yellowish Discharge in Inner Corners of Eye(s) and a Red, Inflamed Conjunctiva

RELATED SYMPTOMS: One or both eyes may be involved. The tearing, mucous, discharge, and redness may be mild to severe. The cat may also blink excessively and/or squint. You may notice a protruding third eyelid, which can be seen at the inner corner of the eye.

POSSIBLE CAUSE: Was your cat recently exposed to a dry wind, either generated by the weather when playing outdoors or while prowling in dusty places? Could she have gotten into your spice cabinet and gotten a bit of pepper, cayenne, or onion powder in her eyes? Does your kitty have hair that hangs in her eyes? Has your pet been diagnosed with a generalized viral infection, such as rhinotracheitis or pneumonitis? Your cat may have **conjunctivitis**, a condition that literally means **an inflammation of the conjunctiva**. The conjunctiva is the normally light pink mucous membrane that surrounds the eyeball. It can be irritated by a number of things, including dry air or wind, a dust particle lodged in the eye, or even a virus that has hit the rest of the body. Allergies are another common cause of conjunctivitis.

CARE: If the conjunctivitis is mild, you may be able to treat it at home. First, address the cause of the condition. For instance, keep your pet indoors or in a sheltered area outside when the weather is dusty or windy; keep soap away from your pet's eyes when bathing her; clip facial hair to keep it out of her eyes; and so on.

Clear away discharge 2 or 3 times a day using a soft cloth dipped in lukewarm water, weak chamomile tea, or a dilute boric-acid solution designed for opthalmic use (which you can purchase from your local drugstore). Then flush the eye with Visine, Murine, or artificial tears (such as methylcellulose) to help remove foreign debris and to lubricate the cornea in order to make your pet more comfortable. Covering your cat's eyes with a damp, warm compress can also be very soothing. Move your cat to a different location when you dust, vacuum, or cut the grass.

If the condition doesn't improve after 1 day of homecare, or if the conjunctivitis is severe, see your vet, who can usually diagnose the condition after a physical exam and a few simple tests. If something is stuck in the eye, your vet will remove it. As for homecare, your vet will recommend that you keep the eye clean and give you an antibiotic and/or anti-inflammatory drops or ointment to apply to the eye 3 or 4 times a day. If the cause is an allergy, cortisone and/or antihistamines may be dispensed.

PREVENTION: Keep hair, soap, and foreign objects away from your cat's eyes. Don't expose your pet to dry wind or dusty environments, such as the basement. If possible, avoid those things to which your pet is allergic. Keep your cat away from other cats who could transmit highly contagious viral infections.

Excessive Tearing with Sticky, Yellowish Discharge in the Inner Corners of the Eyes; Lower Eyelids Are Rolled In Toward the Eye

RELATED SYMPTOMS: Both eyes are involved, and the cornea of each eye is mildly to severely red and inflamed. Tearing, too, may be mild to severe. The cat may also blink excessively and/or squint.

POSSIBLE CAUSE: Is your cat a Persian or Himalayan? Your kitty may have **entropion**. In nonvet-speak, entropion means the eye slit is too narrow to hold the eyeball in place. To give the eye a bit more room, the lower eyelid rolls inward.

CARE: Take your pet to the vet for a diagnosis. Surgical correction of the eyelid deformity is the only permanent cure. In the meantime, if any infection arises, your vet will check for corneal abrasions and treat the infection by cleaning the area and giving you antibiotic drops or ointment to place in the eye 3 or 4 times a day.

If you cannot get to the vet right away, clean the eye several times a day by flushing accumulated discharge and removing any dried discharge from the lids and corner of the eye. You can apply a few drops of artificial tears (such as methylcellulose) to your cat's eyes every 4 hours to offer her temporary relief until the vet sees her.

PREVENTION: There is no known prevention. Because entropion can be passed down to future generations, you may want to think twice before allowing a cat with this illness to breed.

Opaque Film Covering the Inner Corner of the Eyeball, Accompanied by Tearing

RELATED SYMPTOMS: The eyeball may appear to bulge or shrink back into the socket. Blinking or squinting are often present. One or both eyes may be affected.

POSSIBLE CAUSE: Has your pet been diagnosed with a corneal ulcer or glaucoma? Could there be some foreign material in the cat's eye? Any of these conditions can cause the globe of the eye to retract and the third eyelid to be secondarily elevated or protrude up and over the inner corner of the eye.

An abscess or tumor of the eye socket just behind the eye may cause the third eyelid to elevate and become more apparent. Trauma to the eye or simply eye pain can also produce a third-eyelid elevation. A disease infecting the nerve/loss of the nerve stimulation to the third eyelid can cause it to protrude, as can tranquilizers (when administered for an unrelated medical condition). This condition is referred to as **protrusion of the third eyelid**.

CARE: If the third eyelid has protruded because of tranquilizers, it will return to normal once the drugs wear off. Otherwise, take your cat to the vet, who can diagnose the condition after examining the area. If the condition is caused by an infection, antibiotics will be prescribed; if it is caused by an inflammation, cortisone or another anti-inflammatory medication will be given. At home, give your cat antioxidants such as vitamins A, C, and E (see Appendix E: List of Recommended Dosages, pp. 169–189). You also can place 1 or 2 drops of cod liver oil in the affected eye.

PREVENTION: Treat all injuries and infections immediately. Be sure to feed your kitty a healthy diet supplemented with the appropriate vitamins and minerals.

Swollen Red/Pink Lump Sitting Over the Inner Corner of the Eye, Accompanied by Tearing

RELATED SYMPTOMS: One or both eyes may be affected. The cat may also suffer from conjunctivitis.

POSSIBLE CAUSE: Is your pet a Burmese or Persian? There's a chance she may be suffering from a **prolapsed gland of the third eyelid.** This somewhat rare condition occurs when the gland located at the base of the third eyelid (which is believed to play a role in tear production) breaks loose from its attachment to the lid and visibly pokes outward.

The condition is also called **cherry eye** because the prolapsed gland actually looks like one of the small, unformed cherries that you sometimes see piggybacked to a fully formed cherry.

CARE: Take your cat to the vet, who will be able to diagnose the condition after a physical examination of the area. Treatment options include removing part of the gland or forcing the gland to stay put by tacking down the migrating portion to the inner part of the third eyelid. You can keep the eye clean by removing any discharge accumulating in the inner corner of the eye. Flushing the eye with a mild boric-acid solution is also advised.

PREVENTION: There is no known prevention. The condition results from a hereditary weakness.

Head-Shaking and Ear-Scratching

A cat experiencing ear pain can't talk to you about her discomfort. She can, however, let you know that something is wrong by shaking her head (in an attempt to propel the ache and any ear discharge or debris from the ear canal) or scratching her ear(s).

If you see your kitty indulging in one of these behaviors, give her ear a look. Can you see anything caught in the canal? Is there any type of discharge or colored earwax? Is there a foul smell? Can you see tiny, moving, white, flecklike mites? Is there redness and swelling? Has the ear flap ballooned out and filled with fluid? If so, read on. Some of these conditions can be addressed at home, but others require the care of a veterinary professional.

Head-Shaking, Scratching of the Ear, and Tilting of the Head

RELATED SYMPTOMS: If her discomfort is mild, your cat may want her ears scratched more than usual. However, if the pain is severe, she may shrink from having her ears touched. The ear may smell foul and release a thick, yellowish or brownish discharge. The cat may show a loss of coordination, stumble, circle to one side while walking, appear lethargic, and exhibit a marked loss of hearing.

POSSIBLE CAUSE: Has your cat recently received an ear injury? Could there be a foreign body lodged in her cat's ear, perhaps a foxtail that lodged there during an outdoor jaunt? Is your kitty a longhaired breed or breed mix with thick hair growing in the ears? Do you bathe your cat weekly? Could she have an ear tick? (These parasites that are especially common in the Southwest.) Any of these can indicate an **infection of the outer, middle,** or **inner ear.**

CARE: If the pain and discharge are mild, and coordination and hearing seem normal, you can attempt homecare. First, look in the ear for a foreign object, which is most likely to be a foxtail or other plant matter. If you see an object in the opening of the ear canal, you can grasp it with your *clean* fingers or a pair of tweezers and remove it. Once the object has been extracted, clean the ear by filling the canal with an earwax solvent. Massage the ear vigorously and allow the cat to shake her head. Then use a Q-Tip or cotton ball to remove the debris that has been shaken to the surface. Once cleaned, isopropyl alcohol or 3% hydrogen peroxide may be used to disinfect the ear (which may sting if there are abrasions or raw areas in the ear). If the cat seems pained and resists homecare, or if you are unable to remove the object yourself, take her to the vet.

The vet will thoroughly examine the ear using an otoscope (the same device doctors use to look in humans' ears). If the cat is in pain or won't stay still during the exam, she may need to be anesthetized in order to examine the ear canal thoroughly. If the ear contains fluid, it will be flushed and suctioned out at this time. Any underlying cause, such as mites, a tumor, or a lodged object, will be addressed. Antibiotics will be prescribed to clear up any infection. Daily cleaning, application of ear-drops 2 times a day, and once-weekly visits to the vet are necessary until the ear is totally healed. If, in spite of this care, the ear remains infected, an ear culture and antibiotic sensitivity test may need to be performed. On rare occasions, surgery may be needed to establish drainage.

PREVENTION: Thoroughly check your cat's ears weekly and clean them with warm almond oil or an earwax solvent. A common earwax solvent called Oticlens is available from your vet.

Vigorous Scratching of the Ears and Reddish-Brown to Black Earwax

RELATED SYMPTOMS: The ears may also be inflamed and the cat may frequently shake her head. Abrasions on the cat's ear flap caused by excessive scratching may also be present.

POSSIBLE CAUSE: Does your cat spend time with neighborhood felines or canines? Is she a newly adopted kitten? Your kitty may have **ear mites**—tiny spiderlike parasites that are transmitted by direct contact with infected animals and are quite common in kittens who still live with—or have just left—the litter. Ear mites provoke intense itching, which a cat reacts to with equally intense scratching, which, in turn, can lead to ruptured blood vessels. The ear reacts to these mites by producing a dark, viscous, waxy discharge that can clog the ear canal. Both conditions can bring about a secondary infection.

To view the bugs, remove some of the wax from the ear canal with a cotton swab. If you hold the swab up to a bright light or smear the material on a piece of black paper, the mites will look like pinpoint-sized white specks (a magnifying glass is also handy for this). (Note that identifying mites can be difficult, and easily confused with a yeast condition.)

CARE: You have two options: Take your pet to the vet or attempt to treat the mites with homecare. If you go to the vet, he will physically examine the area and study the wax microscopically to determine the presence of mites. If the ear is badly congested with wax, your vet will thoroughly clean the area. A topical ear-mite formula will be prescribed, which you must administer to your pet daily, usually over the course of 1 month. Wash all bedding and vacuum rugs and furniture.

To treat your cat's ear mites at home, you must first clean the infected ear. Fill the ear canal with Oticlens or warm almond oil (available at health-food stores) to loosen wax and debris that have accumulated, then massage the side of the face just below the ear. After allowing several hours for the wax to soften, fill a large plastic eyedropper with equal parts lukewarm water and white vinegar and repeatedly flush the loosened wax from the canal. You can use cotton balls or Q-Tips to remove the debris that rises to the surface.

When the ear seems free of all debris, apply 6 to 10 drops of mineral oil to the ear canal: This will smother the mites. You should continue to clean the ear and apply the mineral oil treatment daily for at least 1 month, when eggs may be hatching. However, if you do not see any improvement after the first 2 weeks, see your vet.

PREVENTION: Whenever you bring a new pet feline home, have your vet check her ears for mites. Furthermore, if your cat has access to the outdoors (where she could come in contact with other cats), have her ears checked during her annual veterinary checkup. If one cat in a multi-cat household has mites, all of the cats should be treated.

Violent Head-Shaking and Swelling of One or Both Ear Flaps

RELATED SYMPTOMS: The cat may also scratch her ear frequently.

POSSIBLE CAUSE: Has your cat recently suffered an injury or other trauma to the ear? It is possible that she has an **aural hematoma**, known in layperson's terms as a huge **blood blister** on the ear flap. Though somewhat rare in cats, the condition occurs when a trauma breaks blood vessels under the skin of the ear, causing blood to accumulate between the skin and the ear cartilage.

CARE: If left untreated, a swollen ear flap can cause severe scarring and shriveling of the ear (cauliflower ear), so it's important to take your pet to the vet. In the event that you cannot get to the vet right away, talk softly and gently to your cat and use any other methods to keep her calm in order to prevent repetitive head-shaking or ear-scratching that can further damage the ear. (You can fashion a makeshift Elizabethan collar out of posterboard to keep your cat from reaching her ear until you get to the vet.) A physical exam is usually enough to diagnose the problem. To treat the condition, your vet will drain the fluid from the ear, then bandage the area in such a way that the skin will be sandwiched to the cartilage. The vet may choose to use an alternative approach that involves lancing and draining the hematoma, then placing numerous sutures from the skin on one side of the ear to skin on the other side, literally sandwiching the separated tissue together until it heals. After about 3 weeks, sutures are removed. Meanwhile, to protect the sutures, your pet probably should wear an Elizabethan collar-especially when you can't be present to make sure she won't pull on them.

PREVENTION: Address all ear injuries immediately. Clean your cat's ears weekly to help prevent ear infections and violent head-shaking and scratching, which can cause the blood vessels to break.

Nasal Sores and Nosebleeds

You probably already know a healthy cat's nose is usually cool and slightly moist, with thick, even, intact skin. Should you notice a sore or scrape on the front of your kitty's proboscis, there's a good chance it was caused by something your cat did, such as rooting around a rough surface. Some autoimmune illnesses can cause a blisterlike rash on the nose or weeping sores around the nostrils.

As for nosebleeds, causes include a **foreign object lodged in the nostril, a blood platelet disorder,** and **cancer.** The blood may flow in a thin stream or may be mixed with mucous. Unless you actually see something lodged in your cat's nose and can easily remove it (see the section, Blood-Flecked Discharge or Pure Blood Discharging from One Nostril, Accompanied by Bouts of Sneezing, p. 51), your best course of action is a trip to the vet for diagnosis and treatment.

Abrasions and Scabs on the Front of the Nose

RELATED SYMPTOM: The nose may be drier than normal.

POSSIBLE CAUSE: Has your kitty been rooting in rough soil with her nose? Has she been pushing her nose through the bars of her cage or your yard's cyclone fence? A "yes" to either of these questions points to a **superficially wounded nose.**

CARE: If oozing sores are noticeable around the nostrils (which can indicate an autoimmune disorder; see Chapter 6), a medical condition may be present: See your vet for a diagnosis and treatment. Otherwise, you can treat your cat's cuts and abrasions at home. To prevent scabs and/or abraded skin from drying out, rub a gentle oil (baby oil or massage oil will work) into the skin of the nose. If you fail to see an improvement in 3 days, see a vet, who may dispense an antibiotic ointment.

PREVENTION: Limit your cat's nose-burrowing activities and attempt to teach her not to rub her nose on rough carpet or yard fencing.

Blood-Flecked Discharge or Pure Blood Discharging from One Nostril, Accompanied by Bouts of Sneezing

RELATED SYMPTOMS: The sneezes are often quite violent and recur in bouts of 3 or more.

POSSIBLE CAUSE: Does your cat play outdoors unsupervised? Are there foxtails and wild grasses, such as wheat, in the vicinity? It's possible that your cat has a **frond tip** of one of these plants **caught in her nose**.

CARE: If you can see the object protruding from her nose, gently and slowly try to pull it out. If the object is not clearly visible, try using a flashlight to look up into her nose. You can use a small pair of tweezers to gently grab any foreign object and carefully remove it from the nostril, taking care not to pinch any normal healthy tissue. If the object won't easily budge or if you are met with resistance once it is partially out, let your vet remove it.

If you can't see the object, it may be lodged deeper in the nasal cavity. Take your cat to the vet, who will thoroughly examine the area using an endoscope to see the offending body. In order to remove the culprit, your vet must anesthetize your cat to prevent her from sneezing or struggling.

PREVENTION: Limit your pet's exposure to barb-ended grasses by keeping her indoors or by reducing the weeds and foxtails on your property.

Bloody Nasal Discharge with Noisy Breathing

RELATED SYMPTOMS: Your cat may sneeze frequently and the nose itself may or may not appear swollen or misshapen. The blood may come from one or both nostrils.

POSSIBLE CAUSE: Is your cat older than four years? There's a chance she could have a **cancerous nasal tumor**, a condition that usually affects older cats.

CARE: Take your pet to the vet, who will give her a thorough physical to reach a diagnosis. In order to confirm the existence of a nasal tumor, your vet will anesthetize your pet so that X rays can be taken and an endoscopic exam performed. If necessary, the vet may use an endoscope to remove tissue from the nostril, and he may perform a biopsy. Should cancer be present, your vet may remove the tumor, although surgical excision may be very difficult in this location. He will then immediately place your cat on radiation therapy. Meanwhile, upgrade your cat's nutrition by adding lightly cooked meat to a diet comprised of high-quality chemical-free, commercial cat food, supplemented with these antioxidants: vitamins A, C, E and beta-carotene (see Appendix E: List of Recommended Dosages, pp. 169–189).

PREVENTION: Pay close attention to changes in your cat's nose and nasal secretions. Nasal cancer cannot be prevented, but the earlier it is caught and treated, the longer and more comfortably your cat will live.

Nosebleed, Accompanied by Blood in the Urine

RELATED SYMPTOMS: You may notice dark pigmented spots on the skin and mucous membranes. The cat's stool may be dark due to the presence of blood.

POSSIBLE CAUSE: Has your pet recently been diagnosed with lupus, a disease partially characterized by skin lesions? Answering "yes" may indicate a decrease in blood platelets. Known as **autoimmune thrombocytopenia**, this is an uncommon condition in which the body destroys its own blood platelets, which are needed in order for blood to clot. As a result, capillaries hemorrhage more readily, and the leaking blood accumulates under the skin and mucous membranes.

CARE: Take your pet to the vet, who can diagnose the disorder after running blood tests. If her condition is mild, your pet may need to be put on corticosteroid therapy, which regulates platelet production. In more severe cases, your pet also may require a blood and/or plasma transfusion to increase the number of blood platelets. To help your cat at home, give her these antioxidants: vitamins A, C, E, and beta-carotene and trace minerals. Providing your kitty with flaxseed oil is also recommended (see Appendix E: List of Recommended Dosages, pp. 169–189).

PREVENTION: There is no prevention.

CHAPTER 5

Mouth and Throat

An affection-filled lickdown, a kitten lapping up spilled milk, a house cat proudly toting her latest catch, a fang-baring snarl given by a spooked cat—a lot of the images we have of kitty-like behavior involve the feline mouth.

A good rule of thumb when considering your cat's oral health is that many of the illnesses and symptoms that apply to you as a human also can affect your pet. These include bad breath due to a kidney disorder or diabetes, problems with tooth enamel, periodontal disease, excess tartar, and an object lodged in the throat. Of course, you examine your mouth and teeth every time you brush your teeth. Although it doesn't hurt to regularly examine your cat's mouth and throat—even to schedule yearly dental cleanings—you also must be aware of signs that signal problems.

Bad Breath

Extremely bad breath may signal one of several things: excessive **tartar buildup** on the teeth, an **infection**, or a **tumor** somewhere in the mouth or throat. Bad breath can also be a result of certain types of food, **digestive problems, uremic poisoning from kidney disease, ketone breath from untreated diabetes,** and other metabolic problems. How can you tell the difference between healthily pungent feline breath and the foul variety? There's only one way: Know the scent of your pet's regular breath by smelling it regularly.

Bad Breath and Excessive Salivation, Possibly Accompanied by Oral Bleeding

RELATED SYMPTOMS: Facial deformities may be present, and the cat may have difficulty swallowing and chewing.

POSSIBLE CAUSE: Is yours an older cat? Is he a short-nosed breed, such as a Persian, Burmese, or Himalayan? If you answered "yes," your cat might have an **oral tumor.** The breeds just mentioned are more susceptible to oral tumors, although other types of cats may also get them. Cancerous oral tumors are most often **squamous cell carcinomas.**

CARE: Take your cat to the vet, who will perform a biopsy to determine what type of tumor your pet has. Some malignant oral tumors are fast-spreading tumors, whereas others remain only locally invasive (i.e., squamous cell). Tumors can be removed either by conventional surgery or cryosurgery (using subfreezing temperatures). After the growth is removed, the antioxidants vitamins A, C, E, and beta-carotene, along with a chemical-free high-quality diet, can help prevent a recurrence (see Appendix E: List of Recommended Dosages, pp. 169–189). Benign growths require no treatment unless they hamper your pet's eating, drinking, or breathing.

PREVENTION: Examine your cat's mouth regularly. You can't prevent a tumor from showing up, but you can prevent it from growing larger by taking your cat to the vet at the first sign of trouble. The earlier a malignant tumor is removed, the better the chances for preventing its spread.

Bad Breath, Swollen Gums, and Decreased Appetite

RELATED SYMPTOMS: Gums will be red and painful and may appear receded. They may also bleed. Your cat may stop eating altogether.

POSSIBLE CAUSE: Is your pet over three years old? Is he an Abyssinian or Somali? Does he eat a soft-food diet? Has he been diagnosed with eosinophilic granuloma complex? (See Chapter 6.) If you answered "yes" to one or more of these questions, your kitty may have **periodontal disease.**

The condition begins when plaque forms on the kitty's teeth. Humans brush this plaque away daily with their toothbrushes. Cats, however, must rely on crunchy cat food to get rid of plaque—a poor substitute for brushing. If not removed, this plaque hardens into tartar, which attacks the gums, which may recede and bleed, and teeth, which may become loose and fall out. This makes for painful eating, and the animal may avoid food altogether.

CARE: Take your pet to the vet. If caught early, periodontal disease is easily treated with a dental cleaning not unlike that your dentist gives you. Because the most critical area of tartar buildup lies under the gums in the "periodontal pockets," an anesthetic must be given for removal.

In more advanced cases of periodontal disease, antibiotics will be given to wipe out any bacterial infection. Loose or severely decayed teeth will be removed immediately to prevent the gums from abscessing (see the following section, Bad Breath, with Swelling Below the Eye), to relieve any pain, and to reduce the overall bacterial population in the mouth.

PREVENTION: Provide crunchy food for your kitty to gnaw on. To help slow the progressive buildup of tartar and accompanying bacteria, daily home dental care is necessary. You can use a soft, children's toothbrush or a specially devised finger brush available from your vet. A special gel is also available that can be rubbed onto the cat's gums, eliminating the need for brushing. The gel dissolves and is circulated throughout the mouth by saliva, helping to slow tartar buildup and consequently reducing oral bacteria and toxins. To help strengthen your cat's gums and fight infection, give your cat vitamin C daily (see Appendix E: List of Recommended Dosages, pp. 169–189).

Bad Breath, with Swelling Below the Eye

RELATED SYMPTOMS: The gum below the swelling may be very inflamed and the conjunctiva surrounding the eyeball may be very red.

POSSIBLE CAUSE: Does your cat have a broken or decayed tooth or periodontal disease? If so, he may have an **abscessed tooth**. When a tooth is abscessed, an infection enters the root canal through the break in the tooth, which can spread to the sinus cavity and the entire jawbone if left untreated.

CARE: Take your kitty to the vet, who will perform X rays to determine which root is infected. Depending on the condition's severity, the vet may perform a root canal or remove the tooth. Follow-up care entails a week of antibiotics. Giving your cat vitamin C and coenzyme Q-10 will help his body fight infection and heal the abscess (see Appendix E: List of Recommended Dosages, pp. 169–189).

PREVENTION: Regular brushing at home and professional dental care help to prevent the formation of abscesses.

Difficulty Swallowing

Swallowing uses a combination of muscles, including the tongue, hard and soft palates, pharynx, and esophagus. Should a problem exist with any of these muscles, swallowing can be difficult. In vet-speak, this condition is called **dysphagia**. It can be caused by a neurologic or muscular disease or a foreign body, such as a tumor, abscess, or even a piece of swallowed matter (maybe a twig or piece of bone).

Quite often swallowing trouble is accompanied by regurgitation, drooling, excessive head movements after chewing, a reluctance to eat, and weight loss. Make a note of any accompanying symptoms: Your keen observances might be just what your vet needs to make a diagnosis.

Difficulty Swallowing and Regurgitation of Food

RELATED SYMPTOMS: Regurgitation of food is effortless and spontaneous. In other words, food will "just come up," usually soon after it is eaten. Regurgitation should be distinguished from retching and vomiting, which require much more force and much more violent movements.

POSSIBLE CAUSES: Such a situation usually results from a **neurological disease** that paralyzes the esophagus and causes it to enlarge or balloon out. The condition, which can be inherited (as with Siamese cats) or appear as a secondary symptom of a **neuromuscular disease**, makes it hard for food to reach the stomach.

CARE: Feed your cat water and semisolid food, such as high-quality canned cat food, and watch him for 48 hours. If the condition does not improve, make an appointment with the vet, who will use radiology and/ or endoscopy to help determine the cause. Some afflicted cats improve with time, others need surgery to improve the condition. Until your cat recovers, your vet may tell you to raise your pet's front legs on a table or chair when feeding him so that gravity will help the food reach the stomach.

PREVENTION: There is no known prevention.

Excessive Drooling and Salivating

Saliva—whether in cats, dogs, or humans—softens, digests, and lubricates food. Based on what we learned from Pavlov's dogs, it's not surprising that anything your pet connects with eating—a can being opened or the smell of dinner being prepared (yours or his)—is going to prod his salivary glands to manufacture more of this sticky liquid.

Usually this saliva stays conveniently inside the mouth. There are instances, however, when it doesn't, and the cat drools: Vets call this **ptyalism**. Some cats slobber excessively when in a state of bliss—perhaps when sitting on your lap or having their ears scratched—but drooling also accompanies neurological conditions (such as **motion sickness), gastrointestinal ailments**, and **metabolic disorders** (like **uremia**). Because saliva helps cool the body, kitties who are overheated, due to either the temperature or overexcitement, often salivate heavily, as do animals who have ingested certain medicines or toxins. If a cat licks something that irritates his tongue or oral membranes, excessive salivation may be an attempt to flush the irritant from the mouth.

If the cat is otherwise healthy, has no history of "bliss-related drooling," has not ingested irritating material, and exhibits no other symptoms, keep him calm and cool for 1 to 2 hours and see if the saliva production slows down. If the saliva keeps coming, call your vet for an appointment. If left unchecked, oversalivating can cause dehydration. Be aware that in some cases drooling has nothing to do with excess saliva production but is a side effect of swallowing difficulties.

Drooling, Scratching at the Mouth, Choking Noises, and Restlessness

RELATED SYMPTOMS: Saliva may be tinged with blood and the jaw may appear propped open. The cat may experience difficulty swallowing, exhibit a swollen, bluish tongue, refuse food, and/or cough.

POSSIBLE CAUSES: Is your cat left unattended for periods of time? Some irresistible item may have caught his eye—anything from a rubber band to a piece of discarded dental floss to a toothpick. After playing with the object, your cat might have swallowed it. The **object may** then **have become lodged in his mouth, throat**, or **esophagus.** This could have happened moments earlier or weeks ago.

CARE: Take your cat to the vet. If an object isn't readily visible, an endoscopic exam may help detect the offending culprit—an important precaution because the drooling and swallowing problems that accompany this condition can be easily mistaken for rabies.

If something is lodged in the mouth, throat, or esophagus, the vet will sedate your cat so that the object can be removed. Your vet will carefully examine the mouth for any resulting cuts, bruises, or abrasions. If infection has set in, your cat will be given antibiotics. After you return home from the vet's, feed your cat a soft-food diet for at least 48 hours to prevent irritating any painful areas in the mouth.

Note: If you suspect that your cat swallowed string, thread, or dental floss—or you actually see string or another long object emerging from a cat's hind end—*do not attempt to pull it out!* You risk seriously damaging the cat's intestines. Instead, take your pet immediately to a vet, who will offer the best advice on what to do. In some cases, he may need to perform surgery to remove the object.

PREVENTION: Never give your cat cartilage, vertebrae, or chicken or fish bones to chew on. Be a thorough housekeeper and keep string, yarn, dental floss, rubber bands, paper clips, and other swallowable objects off the floor and out of your cat's reach. Unfortunately, many cats learn how to open cabinets. So, when stashing your sewing supplies, fishing equipment, and other feline health hazards, make sure your stow-spot is kitty-proof.

Tooth Abnormalities

A cat's tooth is comprised of three layers: the soft, interior pulp, nerves, and blood vessels; the dentine, which makes up most of the tooth; and the dense, brittle, white-colored enamel, which coats the tooth. Like normal adult human teeth, normal adult feline teeth should show some degree of whiteness with no jagged or frayed-looking edges. They should be solidly anchored in firm, pink, nonbleeding gums. Obvious discoloration, a lack of uniformity, broken pieces, looseness, or bleeding or receding gums signal a dental problem—which, in turn, signals the need to visit the vet.

Dental cleanings for pets usually require anesthesia, a practice which may sound frightening to pet owners but is necessary to do the job properly. However, presurgical blood work is a common way to ensure that the animal's body systems are in good shape to undergo anesthesia. It can be used to check for dehydration, white blood cell count, and liver and kidney functions. This blood work can begin at any age, depending on what each veterinarian recommends. (I typically begin using presurgical blood work at age 2, administering it more extensively as an animal ages.)

Discolored, Yellowish Teeth

RELATED SYMPTOMS: The teeth will appear otherwise normal. The cat's breath may or may not be bad.

POSSIBLE CAUSE: Is yours an adult cat who was given tetracycline as an adolescent? Is he a kitten whose mother received tetracycline? Does he have a buildup of tartar? Any of these can cause **yellow teeth**.

If tetracycline is to blame, the teeth are simply stained: They are perfectly healthy. If tartar is the culprit, be advised that if the buildup has been allowed to sit for an extended time, the teeth can remain somewhat discolored even after the tartar has been removed. These clean-but-discolored teeth are also healthy.

CARE: With tetracycline-yellowing, there's not much your vet can do other than to assure you that your cat's teeth are healthy and perfectly functional. In the case of tartar buildup, your vet will clean your pet's teeth and show you how to properly care for them at home.

PREVENTION: Avoid giving young cats tetracycline unless absolutely necessary (ask your vet about doxycycline, a newer antibiotic in the tetracycline family that doesn't produce the yellowing). To prevent tartar buildup, brush your kitty's teeth daily or apply a prophylactic gel (your vet will show you how). Your cat's teeth should be cleaned annually by a veterinary professional.

Jagged-Edged Tooth, Possibly Accompanied by Cuts in The Mouth

RELATED SYMPTOMS: A section of the gums may be inflamed and sensitive. The cat may be reluctant to eat.

POSSIBLE CAUSE: Has he recently experienced some kind of trauma to the mouth, such as a blow? Your kitty may have a **broken tooth**.

CARE: Take your cat to the vet. If the break does not involve the central pulp and the tooth is free of decay, your vet may simply file down the jagged edges. However, if the break exposes the central canal or decay is present, your vet may opt to fill the tooth or, if the damage is extensive, perform a feline root canal or extraction.

PREVENTION: Should your cat receive a blow to the mouth or run into an object, check his teeth immediately for damage. Be sure to examine your pet's teeth regularly for abnormalities. A tooth may be broken horizontally (i.e., the tip breaks off) or vertically (i.e., a slice of enamel comes off the side of the tooth).

Single Discolored, Bluish-Gray Tooth

RELATED SYMPTOM: The surrounding gum may be inflamed and/or painful.

POSSIBLE CAUSE: Do you seldom or never brush your pet's teeth? Has it been more than a year since his last dental checkup? Your cat may have a **decayed** or **devitalized** (dead) **tooth**.

CARE: Take your cat to the vet, who will determine if the tooth is dead, if there is a cavity, or if the blood supply to the tooth has been damaged and the extent of that damage. Depending on the situation, your veterinarian will either extract the tooth or treat it with a filling, root canal, or crown.

PREVENTION: Brush your pet's teeth daily. Do not rely on dry food to keep your cat's teeth clean. Annual to semiannual dental cleanings will help prevent cavities and periodontal disease.

CHAPTER 6

Hair and Skin

Dense, shiny, well-conditioned fur and soft, pliant skin are hallmarks of good health for a variety of mammals, cats included. But a beautiful coat counts for more than just decoration; it protects your kitty's precious internal organs from the environment and helps her body maintain a constant temperature.

Their highly visible status makes feline hair and skin easy-to-monitor indicators of your pet's overall wellness. Because a cat's fur and skin are among the last recipients of the nutrients she digests, if she doesn't receive proper nutrition—or has an illness that leaches whatever nutrients she does ingest—the hair and skin go without their share in order that the precious internal organs can receive nourishment. If this scenario continues, her fur and skin will visibly suffer. Her hair may become thin or fall out in spots altogether; it may grow dull or become greasy. The skin may take on a different hue, turn dry, or become prone to infections—and these are just a few of the possibilities.

Hair Loss

Your cat's coat continuously replaces itself. At any given moment, there are hairs falling out, hairs growing in, and hairs that are resting beneath the surface, waiting for their turn to grow. Unfortunately, you can't determine whether your pet's hair loss is abnormal without first knowing what is normal. Although spring and fall are normal feline shedding seasons, indoor animals and longhaired breeds commonly shed year-round. In other words, if your cat fills a brush each day with shed hair and has always done so, that's normal. (Indeed, if your kitty is a heavy shedder, groom her daily to remove dead hairs and make way for the new ones. This will keep the coat clean and help prevent skin conditions.)

Keep in mind that normal shedding rarely produces patches of exposed skin. Notice bald spots? Suspect a hair-loss culprit other than normal shedding. Abnormal hair loss, which vets call **alopecia,** can be caused by numerous factors, including **infection, ringworm, pregnancy, parasites, nutrition, trauma, stress,** or a **hormonal imbalance.**

General, All-Over Thinning and Scaly Skin

RELATED SYMPTOMS: The cat's fur is lackluster and its color may look faded. Any hair loss is generalized or symmetrical, but usually not patchy.

POSSIBLE CAUSE: Is it possible that your cat has internal parasites? Is your pet's diet of questionable quality? If you answered "yes" to one or both of these questions, the animal may be experiencing **nutrition-related hair loss.**

The old, dead hairs that your cat sheds are continually replaced with a supply of new ones. However, if the diet doesn't provide enough protein and/or other essential nutrients—or has a digestive-enzyme deficiency or poor intestinal absorption—it can't properly form these new hairs. Whether the body wasn't supplied with the proper fuel in the first place or internal parasites are causing the malnutrition, the outcome is sparse, lackluster fur and dehydrated skin.

CARE: Take your cat to the vet. He'll review your pet's diet, check the animal for internal parasites, and arrive at the appropriate strategy. You may want to give your cat mineral supplements containing zinc and sulfur, as well as vitamin B supplements, to further enhance the health of existing hair and to encourage regrowth of the thinning coat (see Appendix E: List of Recommended Dosages, pp. 169–189).

PREVENTION: To ensure your cat grows a thick, lustrous coat, ask your vet to recommend a fresh, high-quality, chemical-free diet that is high in protein and supplies essential fatty acids. Supplement your pet's diet with raw vegetables. Plant-derived digestive enzymes and omega-3 and omega-6 fatty acids (flaxseed oil) added to the animal's diet will help encourage healthy skin and a full hair coat (see Appendix E: List of Recommended Dosages, pp. 169–189). Also, have your pet checked yearly for internal parasites; *immediately* if you suspect something is wrong.

Small, Circular, Gray or Red Bald Patches, Often Accompanied by Scaly Skin at the Center of the Patches

RELATED SYMPTOMS: The bald patches may be more noticeable on the forehead and around the ears and muzzle. The patches may or may not contain draining skin lesions, which provoke licking and scratching. The cat's toenails may appear deformed. The owner may also be infected with red, circular skin lesions.

POSSIBLE CAUSE: Has your cat come in contact with other cats, dogs, and/or pet owners? If not, has she touched soil that neighborhood cats and dogs play in? Is your cat young and poorly nourished? If you answered "yes" to any of these questions, your cat may have contracted **ringworm**. The condition, which happens to be quite common, is not caused by a worm at all, but by any one of several contagious microscopic fungi. The condition's name stems from the round, ringlike sores that often mark the ailment, and the fact that ringworm at one time was thought to be the result of burrowing skin worms.

CARE: Ringworm is contagious. If you suspect your cat has ringworm, try not to touch her until she has visited the vet, who will confirm the condition after performing a fungal culture. (Be sure to keep children—who are especially susceptible to ringworm—from handling affected cats.) Treatment of ringworm includes iodine or chlorhexidine shampoos, lime-sulfur dips, topical antifungal medication, and/or oral medications, such as griseofulvin or ketoconazole. Tea tree oil, available at health-food stores, may be painted on and around bare spots several times each day. Because the disease progressively spreads outward, you must also apply any topical medicine to what appears to be normal, still-healthy skin around the edges of the lesion.

Some humans are more susceptible to ringworm than others, so be very careful when handling your pet—for example, wash your hands often or wear gloves when touching her.

PREVENTION: To stop a recurrence of ringworm, wash and disinfect—or discard—your cat's bedding, collar, leash, sweaters, and grooming equipment. Since untreated ringworm spores can survive in dry environments for up to 4 years, disinfect all hard indoor and outdoor surfaces. Mixing 1 part Clorox bleach to 10 parts of water makes an effective disinfectant that can be mopped and sprayed onto surfaces and used to soak certain washable materials. Fungal spores can also live in your heating and air-conditioning systems, so be sure to change all airfilters. Enhance the cat's nutrition with a high-quality, chemical-free diet and stimulate the immune system with nutritional supplements, including vitamins A, C, and E, beta-carotene, zinc, sulfur, and essential omega-3 and omega-6 fatty acids (see Appendix E: List of Recommended Dosages, pp. 169–189).

Scratching and Licking

Cats scratch for a reason: They itch. But why they itch isn't so readily answered. Many once-in-a-while itches are produced by the very same things that make humans scratch, from bathing too frequently with drying shampoos to exposure to skin-dehydrating central heating. Other reasons for a random itch include direct contact with hot or cold surfaces or a localized area of pain. Of course, licking and scratching also can result from **allergies, skin infections, nutritional deficiencies**, and occasional **hormonal imbalances**.

You should become concerned when scratching is prolonged and/or furious. This scratching may be accompanied by biting or licking, and it often signals an underlying health condition or the presence of external parasites. Be aware that a cat's body takes only so much of this aggression before repaying the assault with reddened skin, scaliness, localized hair loss, and infected sores in the affected areas.

Frequent, Intermittent, Intense Scratching

RELATED SYMPTOMS: The cat may also chew or nip at her skin. You may actually see dark purple-gray skin tags, mahogany-colored scurrying bugs, or white flecks (called nits) clinging to individual hairs. Dry patches of skin or pimplelike sores may appear. The skin is often red and traumatized by intense biting and scratching.

POSSIBLE CAUSES: Does your cat spend time outside? Is she in the presence of other animals? Do you board your animal? She may be playing host to **fleas** or other external parasites, such as **ticks** or **lice**. External parasites are often visible on an infected animal: Ticks can appear as plump, dark tags firmly rooted in the cat's skin, and they are sometimes mistaken for skin tumors; fleas are fast-moving, dark brown, oval-shaped creatures that scurry across the skin, leaving tiny, black, comma-shaped droppings wherever they defecate; and lice may look like small, black dots attached to separate hairs.

These pests feed by biting the animal and sucking her blood. (Although humans can get lice and ticks, the parasites are usually not transmitted from cat to human. Although people are bitten by fleas—and in some cases are allergic to them—they never become flea-infested.) In addition to being unpleasant, parasites cause skin ailments. The parasite bites the cat, and the cat scratches—often so furiously that the skin becomes broken and prone to secondary infections.

Here's what else external parasites can do: These bloodsuckers can transmit germs and diseases—for instance, **Rocky Mountain Spotted Fever** and **Lyme disease** travel via ticks—into the bloodstream. A heavy infestation can cause anemia, which can be fatal if left unchecked. Also, some kitties are allergic to parasites, especially fleas: One nip alone can prompt a serious skin problem, such as **flea bite allergic dermatitis**. Fleas are also notorious for transmitting tapeworms.

CARE: Suspect parasites? Thoroughly check your cat's fur and skin for ticks, fleas, and lice—and for the droppings of these parasites. (Combing the cat with a flea comb can help identify parasites and possibly even remove them if they have not become attached.) Should you find your cat is infested, ask your vet to recommend a relatively nontoxic insecticide. Incorrect use of many commercial insecticides can lead to poisoning, so you should thoroughly read the product's label. (Most insecticides call for weekly to monthly use.) Many natural nontoxic insecticidal products are available; you would be wise to try these before using the more toxic chemicals. Try to bathe your cat every 3 days with d-limonene shampoo (found at a pet store or your vet's office) and/or obtain a once-a-month flea and tick preventive (ask your veterinarian for a recommendation). (Be aware that care may differ if your pet is a kitten.)

Treating the house and yard for parasites, especially fleas, is very important in the total elimination of these problems. Many types of flea foggers and housesprays are available, but a safer alternative is using a nontoxic borate powder that can be applied to rugs after vacuuming.

If a tick is the culprit, remove it immediately, according to the following instructions. Generously soak the tick with rubbing alcohol or a nontoxic insecticide, being careful not to get the liquid into your cat's mouth, nose, eyes, or ear canals. Wait 5 minutes. (Avoid touching the tick: Direct contact with the bug may increase your chances of contracting a disease from it.) With tweezers, grasp the tick where it emerges from the skin and pull it out with slow, steady pressure. Jerking and twisting movements can break the parasite's body into pieces, increasing the risk that you'll leave a bit of the bug buried in your cat, which, in turn, can lead to inflammation and infection. Check to see that the entire tick has been removed (you can see a bump in the skin if it is still there). If the tick is gone, wash the area with antibacterial soap and water and dab it with rubbing alcohol to cleanse it. If the tick has not been removed, call the vet, who will dig the tick out.

Don't limit treatment to the obviously infested cat. Any other pets in the home should be treated as if they, too, have external parasites. Clean all animal bedding every few days and don't forget to treat your house and yard. If your house is at least 40 percent carpeted, a borate powder made specifically for cleaning rugs can be found at a pet store or your vet's clinic and can be easily applied: The powder is very safe and will provide up to 1 year's protection against fleas. Yes, these pests infest even the cleanest home. Fleas, for instance, spend only about 10 percent of their time actually on the animal. Therefore, if you treat the cat and not her surroundings, the pests will persist.

PREVENTION: Limit your pet's contact with strange animals, and check your kitty and your surroundings weekly for parasites. Immediately treat any infestations. Feeding your cat a high-quality, chemical-free diet and vitamin-mineral fatty-acid supplements (your vet can recommend a single, complete supplement) can help ward off parasites, who tend to seek out animals with weaker immune systems.

Intense Itching with Redness and Swelling of a Specific Area

RELATED SYMPTOMS: Signs of redness and swelling are usually found on relatively hairless spots such as the underside of the chest, abdomen, and feet; however, they can appear, although infrequently, anywhere on the body.

POSSIBLE CAUSE: Within the last 24 to 72 hours, has your pet come into physical contact with a "new" chemical substance for the first time, including (but not limited to) a shampoo, pet spray, insecticide, perfume, paint, or household solvent? Do you have new carpeting or have your sprayed your lawn with chemicals? Have you recently washed or treated your pet's bedding with a new product? If you answered "yes" to any of these questions, your pet may have a **contact allergy**. A cat's fur acts as an armor to shield the body from contact allergens. Areas where hair is naturally thinner have less protection, thus they are more easily affected by irritants.

CARE: First, give your cat a good bath with a hypoallergenic pet shampoo. Then make a thorough list of all the new chemicals with which your cat has recently come in contact. Take this list and your pet to the vet, who will work to identify the allergy-producing agent. In addition to asking you to remove the offending agent, your vet will administer an anti-inflammatory drug. For your pet's relief, an injection, oral medication, and skin lotion may be required.

If you cannot get to the vet right away, there are several things you can do to treat your pet at home. Bathe her with a shampoo containing colloidal oatmeal (available at pet stores) to alleviate scratching. You can apply a lotion containing aloe vera and chamomile to the inflamed sites. Also, give the cat a chlorpheniramine (Chlor-Trimeton is a recommended brand) or Benadryl antihistamine tablet to relieve further itching. Vitamin C, a natural antihistamine, can also be used with either antihistimine given (see Appendix E: List of Recommended Dosages, pp. 169–189).

PREVENTION: Keep your cat away from those substances identified as allergens.

Obsessive Licking, and Perhaps Scratching, of a Specific Area

RELATED SYMPTOMS: If the licking has been going on for more than 3 or 4 days, look for a vaguely oval, firm, plaquelike ulcer amidst a bald patch. The affected area may also appear red to yellow-brown (stained by the saliva). The ulcer (the hairless spot) and the staining occur only in the area of licking, which can be located on the lips, torso, or inner part of a back leg. In some cases, the cat's lip may be swollen and appear to have been gouged out.

POSSIBLE CAUSES: Is your cat a female under six years of age? She may have **eosinophilic ulcers** (also known as **rodent ulcers**), or she could have **linear granulomas**. The two related conditions are known collectively as **eosinophilic granuloma complex**. Unfortunately, the cause of the ailment is unknown.

CARE: Take your pet to the vet, who may perform a skin biopsy, study the animal's history, and note the ailment's clinical features in order to rule out other diseases. The most difficult part of treating this condition is stopping the cat from licking and scratching.

Treatment typically involves administering corticosteroids orally or via injection for up to 1 month. Should neither of these methods work, radiation therapy may be given to control the lesions. To help your cat at home, try applying a lotion consisting of aloe vera, chamomile tea, and tea tree oil (available at health-food stores) to the affected area. To discourage your cat from licking the affected area, you can apply Bitter Apple, a product that has a terrible taste in order to reduce the cat's tendency to lick. You might even put an Elizabethan collar around the cat's neck to mechanically prevent licking of certain areas until they have healed.

PREVENTION: Because the cause of eosinophilic granuloma complex is unknown, the condition is difficult to prevent.

Scratching and Biting of the Coat, Accompanied by Reddened Skin

RELATED SYMPTOMS: The reddened skin may feel abnormally warm to the touch. You may also notice small bumps, oozing areas, scabs, and/ or dandrufflike scales. Areas where scratching is severe may become infected. Head-shaking and ear-scratching are also very common. Watery nasal discharge, sneezing, and tearing may occur.

POSSIBLE CAUSE: Does your cat have a known—or possibly unknown—allergy to a specific food, insect, chemical, additive, or airborne matter? Her skin may be affected by a condition called **allergic dermatitis**. Some cats—just like people—develop reactions when exposed to certain substances in their environment.

Exposure to allergens can occur through inhalation (called atopy), ingestion, inoculation, or insect bites, or direct contact with the irritating substance. If the cat is allowed to continue severely scratching, she can develop hair loss and a thickening of the skin. (Note: Because contact allergies are discussed in the section, Intense Itching with Redness and Swelling of a Specific Area, this section focuses on atopy and food allergies.)

CARE: Take your cat to the vet. To determine whether an allergy is responsible for your pet's condition (and what that allergy is), your vet will run several tests, including skin and blood evaluations. Skin testing involves injecting small amounts of common allergens under your cat's skin to note the animal's reactions. Blood tests, such as the RAST (radioallergosorbent test) or ELISA (enzyme-linked immunosorbent assay) test, may be used in place of skin testing. Special elimination diets— usually lasting 4 to 8 weeks—can help pinpoint and treat food-related sensitivities.

Your vet may recommend baths 1 to 2 times a week with a gentle hypoallergenic soap to help relieve skin inflammation and prevent secondary bacterial infection. Antioxidant therapy, especially with sulfur and zinc, can help minimize symptoms. Vitamins A, C, and E and beta-carotene also are helpful (see Appendix E: List of Recommended Dosages, pp. 169–189).

PREVENTION: Once you discover the guilty substance, keep it out of your cat's reach. If it is not a substance that can be eliminated from the food or the environment, consider starting your pet on a desensitizing series of injections.

These diets use a single protein source, such as lamb, mutton, or venison, and a single carbohydrate source, such as rice or potato. Any diet should be free of all chemical preservatives, coloring agents, and flavoring agents.

Skin Abnormalities

True, your cat can't tell you what's ailing her, but that doesn't mean there aren't ways for you to tell. The condition of your pet's skin, for instance, says a great deal about her general state of health. A healthy animal has smooth, pliable skin. Her skin will have no excessive scaling, scabs, foul-smelling secretions, or parasites. Depending on the breed—or mix of breeds—of your cat, her skin will range from pale pink to medium brown to black. She may even have spotted skin.

Once a week, take a thorough look at your cat's skin, also known as **epidermis.** If you discover any unexplained changes, including sores, flakiness, abraded spots, or any type of growth, call your vet. Many skin problems are easily explainable: Maybe your cat was scratched by another neighborhood cat or tore a bit of skin trying to clear a fence. Other conditions can be caused by anything from a bacterial infection to a change in diet or environment to parasites. Some conditions can even be an external reflection of an internal disease.

Ball-Like Lump On or Under the Skin

RELATED SYMPTOMS: One or more of these masses may be present, most likely located on the head, neck, or back. A single growth can range from pea-sized to bigger than a golf ball, and it can be moved with the skin.

POSSIBLE CAUSES: Your cat might have one of several types of **skin or sebaceous gland enlargements** or **tumors**. Skin growths are usually above the skin, whereas sebaceous and fatty growths, i.e. **lipomas**, are under the skin. All these growths are quite common among adult cats and are usually benign, but not always.

CARE: Because you have no way of knowing what type of tumor your cat has or whether it's benign or malignant, your pet must visit the vet. He may lance, aspirate, or remove the mass and analyze its content by sending the growth to a pathologist, who will make microscopic slides of the growth and report back to the vet. If it is a cyst or a tumor, the slides will reveal whether it is malignant or benign. If the lump turns out to be a sebaceous cyst, it is usually lanced, cleaned, and chemically cauterized rather than surgically removed. (Antibiotics are infused at the time of lancing, and generally are not required as part of follow-up care.) Homecare includes cleaning the incision 2 times a day, then returning to the vet in 10 days to have the stitches removed.

PREVENTION: Although you can't prevent tumors, you can keep them from growing worse. See a vet upon discovering one, and *do not squeeze* or try to "pop" the lump (it isn't a pimple): Local pressure can irritate and inflame the skin.

Broken, Reddened Skin, Accompanied by Pimples, Pustules, and/or Dry, Crusty Patches

RELATED SYMPTOMS: Your cat's skin might be weeping, and you may notice flakiness. She may have mild to severe scratching and spots of hair loss. You may notice small pimples: These pimples may actually be small pustules that break open when scratched.

POSSIBLE CAUSE: Has your cat recently been treated for external parasites? Does she come in regular contact with other felines? Does she have a hormonal or immune-system malfunction? Is her diet unbalanced? If your cat fits into any one of these seemingly unrelated categories, she could have a type of **bacterial skin disease**.

Cats aren't usually struck by bacterial skin diseases unless they have some underlying health disorder that lowers their skin defenses. When this fortification goes down, the disease-causing bacteria found naturally in the environment seize the chance to play house on your cat's skin. Many of these diseases have human counterparts.

CARE: Bacterial skin conditions are usually limited to the skin's outermost layers. If left untreated, however, they can spread to the deeper layers, making the condition more serious and treatment more difficult. It's important to take your animal to the vet, who will prescribe oral antibiotics, immune stimulants, and nutritional supplements.

To promote healing, keep skin lesions clean and dry. Your vet may prescribe daily administration of oral antibiotics and/or antibiotic ointment and baths (1 to 2 times a week) with medicated shampoo.

If you would like to try to treat a bacterial skin disease at home before going to the vet or you are unable to get to the vet right away, you can apply soothing, anti-inflammatory, antibacterial sprays (containing tea tree oil, aloe vera, and chamomile—available at health-food stores) to the most inflamed areas several times a day. Bathing your cat 2 times a week with a sulfur-based shampoo, immediately followed by an oatmeal rinse containing moisturizers (both available at pet stores or your veterinary clinic), is the best way to treat a generalized bacterial infection that affects many areas of the body.

Add vitamins C and E as well as zinc and sulfur to your cat's diet. Proteolytic plant-enzyme dietary supplements strengthen the immune system and are very helpful (see Appendix E: List of Recommended Dosages, pp. 169–189). If an allergy is the underlying cause of the bacterial infection, you also should switch your pet's diet to a high-quality, chemical-free one with lamb and rice or venison and potato in place of beef.

PREVENTION: Carefully monitoring your pet's health and maintaining it with the proper nutrition will result in a strong immune system that will resist bacterial invasion.

Extremely Dry, Crusty Skin with Grayish-White Flakes and a Dry, Dull Coat or Waxy, Oily, Crusty Skin with Yellowish Flakes and Rancid Odor

RELATED SYMPTOMS: Your cat may exhibit extremely mild to moderate itching. Broken skin and varying degrees of hair loss due to scratching may be present. The outer ear may be inflamed and waxy.

POSSIBLE CAUSE: Does your cat's thyroid function poorly? Does she have allergies or a skin infection? If so, then your pet may have **seborrhea**, a condition that is fairly uncommon in cats. (It is probably more common in dogs and humans.) Marked by an abnormal skin-cell turnover rate, this off-kilter skin-cell production can manifest itself in either of two ways, leading to either excessive dryness or excessive skin-oil production. Seborrhea can show up on its own or be caused by a number of diseases.

CARE: Take your kitty to the vet. Diagnosing seborrhea isn't especially difficult, but determining why the cat has the condition can be. Your vet will run a series of tests (e.g., a thyroid function, bacterial culture, or a skin biopsy) to determine whether seborrhea is your pet's only condition or whether it is a result of another primary illness. If it is caused by another underlying disease, that condition will be addressed first, then your vet will treat the seborrhea.

Treatment of seborrhea usually entails bathing your cat 1 to 2 times a week with medicated shampoos containing chlorhexidine, sulfur, and/ or selenium disulfide (available at pet stores or your vet's office). If the kitty's seborrhea is of the dry variety, your vet may recommend following the bath with a moisturizing skin rinse. He may also prescribe fatty-acid supplements. You may want to add vitamins C and E, zinc, and sulfur to your cat's diet, as well as a plant-enzyme dietary supplement containing high levels of lipase (see Appendix E: List of Recommended Dosages, pp. 169–189).

PREVENTION: When seborrhea is caused by another illness, prompt attention to that condition can prevent seborrhea from becoming more entrenched.

Large, Fluctuant Bump That Feels Warm and Pains the Cat When Touched

RELATED SYMPTOMS: The area is red, swollen, and inflamed. The cat may have a fever, seem depressed, and/or have little interest in food.

POSSIBLE CAUSE: Has your cat been bitten or otherwise wounded recently? If the site has become infected, a **skin abscess** may have developed. Abscesses are formed when bacteria infects a wound, destroying skin tissue and creating a cavity where pus collects. If the cat's immune system is depressed or the invading bacteria are especially hardy, the infection may reach the bloodstream, where it can poison the entire body and eventually lead to death.

CARE: If you are not able to get to the vet for a while, you can try to treat your cat at home. If the bump is an abscess, but is not ready to lance, place a very warm, wet washcloth on the spot several times a day. This helps localize the infection and soften the area around the pus, causing the abscess to "point" and eventually open and drain. Flush the abscess with 3% hydrogen peroxide, and fill the abscess pocket with colloidal silver (available at health-food stores) 3 times a day. Keep the wound open and continue flushing and medicating for at least 4 days. You can supplement your cat's diet with oral proteolytic plant enzymes, zinc, sulfur, vitamin E, and vitamin C (all of which help to support the immune system) (see Appendix E: List of Recommended Dosages, pp. 169–189).

If, after attempting homecare, the abscess is not healing or appears infected, take your cat to the vet. After draining the abscess—or if it has burst before arriving at the vet's—your vet will probe, clean, flush, and medicate the abscess's crater so it can heal. Oral antibiotics are usually prescribed, as well as an antibacterial solution that the owner will use to flush the abscess at home.

PREVENTION: Immediately flush and clean all wounds with mild soap and large amounts of warm, clean water. Hydrogen peroxide (3%) is a good flushing and disinfecting agent. Cat bites are particularly likely to produce abscesses.

CHAPTER 7

Chest, Heart, and Lungs

Your cat's cardiopulmonary system—known as the circulatory and respiratory systems, respectively—is a vast network of blood vessels and major organs, such as the heart and lungs. Because this part of your pet's anatomy so closely resembles your own, you will probably recognize any corresponding symptoms that broadcast your cat is ill.

As they would with a human, problems involving the respiratory system often have very noticeable signs—namely, labored breathing and a surprising variety of coughs. Most of these signals appear suddenly or develop over a few days to a week.

Circulatory conditions can be trickier to diagnose. Many of the conditions come on slowly, and their symptoms—such as the persistent cough of heartworm—may leave you thinking your cat has a cold. Other signs are subtle, including constant fatigue, bluish-tinged tongue and gums, and restlessness. Just how do you learn if your cat has a circulatory condition? Keep a keen eye to any change in your cat's behavior and report any findings to your vet.

Breathing Abnormalities, Including Wheezing and Panting

The purpose of breathing is to provide oxygen for the body and to eliminate the waste gas, carbon dioxide. Upon inhalation (also called inspiration), the diaphragm expands; upon exhalation (also called expiration), it relaxes. Yes, it sounds elementary, but the process is important to observe, since an absence of this diaphragm movement is often a signal that something is wrong.

Your pet's normal breathing is worth monitoring. For a cat at rest, a typical breathing rate is between 20 and 30 breaths per minute. (Each time the chest rises for inhalation then relaxes for exhalation is considered 1 breath.) An increase in your pet's normal rate can be caused by pain, high environmental temperature, fever, fear, nausea, exercise, or excitement. Or, if done with an open mouth and shallow breaths, this increase is called **panting** and is a feline's no-fail method for lowering his body temperature following physical exertion or during hot weather. When accompanied by an anxious look and labored efforts, however, panting can signal **heat stroke.**

None of these fit the mark? An increase in the breathing rate can also signal a **respiratory-tract disease** or a **heart condition.** On the other hand, a severe decrease in a cat's breathing rate is commonly associated with **shock** or a **neuromuscular disease.**

Often a cat's breathing rate is normal, but the animal has difficulty either inhaling or exhaling. Vets call this **dyspnea,** characterized by noisy breathing and/or deep, forceful respiratory efforts. (Of course, the possibility also exists that your cat is both breathing at a different rate *and* having trouble doing so.)

Anatomy could be the cause: Keep in mind that certain breeds— namely flat-faced, short-nosed cats like Persians, Burmese, and Himalayans—are more prone to labored breathing. Noisy breathing is also caused by an obstruction in the nasal passages, mouth, or larynx as the cat vigorously tries to inhale or exhale (depending on the type of obstruction) against the obstruction.

Wheezing is considered a specific type of labored breathing, and it sounds just like the wheeze of an asthma-suffering human. Wheezing is the result of a lung ailment, such as an allergic reaction or infectious bronchitis, that forces a cat's small airways to spasm or constrict.

A cat suffering from breathing difficulties might exhibit other signs. The animal may refuse to lie down and may have an anxious expression and/or an open, gaping mouth. The cat may also salivate and/or extend his head and thrust his tongue out. Check your pet's tongue and gums. Do they show a grayish or bluish discoloration? If so, blood oxygen is low: **Take the cat to a vet right away.** Make every attempt to comfort your pet, and thus slow his breathing and reduce his respiratory efforts.

Breath Accompanied by a Hissing Noise Originating from the Chest

RELATED SYMPTOMS: The hissing noise actually emerges from the chest, not from the mouth or nose. The tongue and gums may be tinged a pale blue or gray.

POSSIBLE CAUSE: Can you see an object lodged in your cat's chest or a piece of rib bone breaking the skin? Is there a puncture wound or gash in his chest area? Your pet may have a **penetrating chest wound,** which created an opening in the chest wall. As air leaks into the chest cavity, the lungs can collapse, making it impossible for the cat to breathe.

CARE: Your goal is to close the wound as quickly as possible. If you have a *clean* cloth or compress nearby, hold it against the wound. If someone else is available, have that person find an Ace-type wrapping bandage that you can wrap over the compress and around the circumference of the chest and across the shoulders.

Another option is closing the wound with your *clean* fingers. Pinch the tissues together to make an airtight seal. If the respiratory distress is severe and the tongue and gums have turned blue or gray, the blood's oxygen level is low. If a vet is not in close proximity, place a small tube—this can even be the *clean* casing of a pen—into the chest cavity before pinching the tissue closed. To keep the lungs from collapsing, it is important to remove as much air from the chest cavity using whatever suction apparatus you have at hand: a syringe, turkey baster, or your mouth can be used to suck air out of the chest through the tube you have placed in the wound. As soon as the suction has been performed, remove the tube and immediately seal the wound. **The cat should be transported to the veterinary hospital as quickly as possible**. Immediate action is necessary to cut the chances that the wound will be fatal.

PREVENTION: Do not let your cat play outdoors unattended.

Difficulty Breathing, Possibly Accompanied by Lameness or Paralysis, Loss of Appetite, and Lethargy

RELATED SYMPTOMS: There may be an accumulation of fluid in the cat's abdomen, signaled by a swollen belly. Collapse may occur in advanced stages.

POSSIBLE CAUSES: Is your cat a male and young to middle-aged? Is your kitty a middle-aged to elderly Siamese, Burmese, or Abyssinian? It is possible he has **cardiomyopathy**, a disease that affects the heart muscle and results in an enlarged heart.

There are two main types of cardiomyopathies: **hypertrophic cardiomyopathy** and **dilated cardiomyopathy**. Hypertrophic cardiomyopathy is the most common kind of heart disease found in cats, usually striking middle-aged male cats. It is characterized by an enlargement of the left ventricular wall, papillary muscles, and septum. The enlargement prevents the heart from expanding properly to receive blood, consequently resulting in blood clots and cardiac arrhythmias. Dilated cardiomyopathy, which occurs most often in middle-aged or older cats and frequently appears in Siamese, Burmese, and Abyssinian breeds, involves the enlargement of all the heart's chambers. This enlargement stretches the heart muscle cells and subsequently creates weaker, thinner heart walls. A lack of taurine in a cat's diet has been proven to be a cause of dilated cardiomyopathy.

CARE: Because cardiomyopathy is a serious, life-threatening condition that can lead to other problems (including thromboembolism of the iliac arteries; see section in Chapter 9, Sudden Lameness and Collapse of Rear Legs, pp. 144–145), **take your cat to the vet right away if you notice these symptoms.** To diagnose cardiomyopathy, your vet may perform an electrocardiogram, a radiography, or blood tests.

If your kitty is diagnosed with hypertrophic cardiomyopathy, your vet may be able to medically control the resulting clots and arrhythmias with diuretics, negative inotropic medication (used to reduce the force of heart muscle contractions), and *carefully monitored* doses of aspirin. (Note: Aspirin is not normally administered to cats. It should be given only under close veterinary supervision.) If your pet is found to have dilated cardiomyopathy, your vet will prescribe taurine, positive inotropic medication to strengthen the heart muscle, and diuretics to remove excess body fluid.

PREVENTION: Be sure to supplement your cat's diet with taurine (see Appendix E: List of Recommended Dosages, pp. 169–189). Have your pet examined by a vet at the first sign of a heart condition.

Pumping Breathing

RELATED SYMPTOMS: There may also be a shortness of breath with the absence of normal breathing noises, as well as an increased pulse rate and pale blue-or gray-tinged tongue and gums.

POSSIBLE CAUSE: Does your cat play outdoors unsupervised? Was he hit by a car recently? Have you seen him take some type of blow to the chest? Should something hit your pet in his rib cage, small tears may develop in either the skin around the rib cage or the interior lung tissue. Air seeps through these tears into the pleural cavity, the space between the lungs and the chest wall. **Pneumothorax** is the condition's official name, and means **air in the pleural cavity.** The pressure of this "wayward" air not only keeps the lungs from fully expanding, it causes them to collapse. As a result, the cat may suffocate.

CARE: Make sure your cat can breathe freely. Sit him up—perhaps supported by a pillow or rolled-up blankets—then rush him to the vet. If the capacity of the lungs is reduced to less than one-third the normal volume, air must be sucked out (see the description under the Care section of the symptom, Breath Accompanied by a Hissing Noise Originating from the Chest p. 76). Once this is done, the lungs can expand again. If the lung is ruptured, your vet will repeatedly remove air from the pleural cavity until the lung wound heals.

Once home, provide fresh air, quiet, and warmth. If a fever develops in the process, antibiotics will be given to ward off infection.

PREVENTION: Supervise your cat's outdoor play. Perhaps this goes without saying, but never, ever kick or throw an object at your pet's chest. (And don't allow any of the neighborhood kids to either!)

Pumping Breathing with Forced Exhalation

RELATED SYMPTOMS: When exhaling, the abdominal wall is pressed unnaturally inward. The following symptoms may also be present: shortness of breath, bluish or gray gums and tongue, vomiting, belching air, and weight loss.

POSSIBLE CAUSE: Has your animal recently been hit by a car, bike, ball, or other object? **Tears in the diaphragm** (also known as a **diaphragmatic hernia**) can result from any strong impact. They also can be present from birth and go undetected for years. If left unchecked, abdominal organs such as the stomach, intestines, spleen, and liver can migrate into the chest cavity. These traveling organs compress the lungs and interfere with breathing, possibly leading to suffocation.

CARE: Visit your vet. A radiograph will help determine whether or not a torn diaphragm is the problem. If it is, the same test will show whether stomach and intestinal organs have settled in the chest cavity. Surgery can correct diaphragm tears.

PREVENTION: Although you can't thwart defects present from birth, keep your cat away from cars and supervise all outdoor play.

Rapid, Shallow Irregular Breathing with Panting

RELATED SYMPTOMS: Your cat might sit or stand with his elbows out, his chest fully expanded, and his head and neck stretched downward in an effort to draw more air into his lungs. He may refuse to lie down. His tongue and gums may be pale and gray- or blue-tinged, and his breathing may be open-mouthed. The cat may be depressed and lethargic, and may exhibit any or all of the following: a weak, suppressed, painful cough; intermittent fever; periods of restlessness followed by exhaustion; and pumping breathing when the animal sits or stands still.

POSSIBLE CAUSES: Could your cat have come in contact with a sick feline? Has a foreign object recently been removed from his trachea? Has your kitty been diagnosed with feline leukemia or feline infectious peritonitis? Has he experienced a blow to the rib cage? He may be experiencing **pleural effusion** or **pyothorax**: In lay terms, pleural effusion indicates a fluid (whether blood, serum, pus, or lymph) buildup in the chest cavity surrounding the lungs; pyothorax is a more narrowly defined condition that indicates pus in the chest cavity.

Often prompted by an interior injury or a bacterial, fungal, or viral infection, fluid escapes into the chest cavity. This substance can be blood from injured blood vessels, lymph from a severed lymphatic vessel, or pus from an infection. It can also begin as blood or lymph and develop into pus as the space becomes infected. This liquid surrounds the lungs and keeps them from expanding, causing the cat to develop respiratory difficulties, eventually resulting in suffocation.

CARE: Any excitement can worsen the symptoms, potentially throwing the cat into fatal respiratory arrest, so handle him carefully and gently. **Take the animal to the vet immediately.** She will investigate the cause with X rays, blood tests, or by drawing off a sample of the chest fluid. Although specific treatment of pleural effusion and pyothorax is based on the underlying cause of the fluid accumulation, draining your cat's chest will probably be necessary. Regardless of what prompted the condition, follow-up care includes about 3 days of hospitalization/observation and antibiotics. A quiet, stress-free environment is very important during the first 3 weeks of recuperation. Provide your pet with a superior, high-quality, chemical-free diet, supplemented with antioxidants such as vitamins A and C (see Appendix E: List of Recommended Dosages, pp. 169–189).

PREVENTION: Supervise all outdoor play.

Vigorous Breathing Efforts with No Chest Movement

RELATED SYMPTOMS: The gums and tongue are tinged a pale blue or gray, and the cat appears to be losing—or has already lost—consciousness.

POSSIBLE CAUSE: If your cat has been out of your sight, he could have **swallowed or inhaled something that is completely obstructing his upper airway.**

CARE: If your cat is conscious, perform a modified Heimlich maneuver: Lie him on his side—preferably on a table—and position your hands around his rib cage just behind his last rib. Give 4 quick, forceful, slightly upward thrusts. This movement forces the diaphragm up while producing an exhalation of air that is usually forceful enough to dislodge whatever is stuck. Peer inside his mouth to see if the object has been dislodged. If not, repeat the thrusts.

Hopefully the animal will not lose consciousness but if he does, you have 60 to 120 seconds to examine the back of his mouth and throat before his heart stops beating. In the event that his heart stops and you are miles from any professional help, you can administer CPR. Start by opening his mouth and laying the tongue to one side between the top and bottom molars. Place your finger in the cat's mouth and throat to feel for any obstructions, including vomit, mucous, or blood. Remove anything you find. *Extend the cat's head and neck, then close his mouth. Inhale deeply, completely cover your pet's nose and mouth with your mouth, then exhale into his nostrils.* The air should reach his chest: Watch for the chest to swell. Remove your mouth and allow the cat's chest to deflate normally. When it has, put your mouth over his nose and mouth and start again. This inflate-deflate cycle should be done 12 times per minute, until the cat begins breathing on his own.

Often, immediately after a cat has ceased breathing (or just prior to it), his heart may stop. (You can check for his heartbeat by wrapping your hand around his chest, just behind his front legs, and applying slight pressure. Alternately, you can check the cat's pulse, which is best felt on the inner side of the thigh in the groin area.) If the nearest vet is some distance away, you also will have to perform external cardiac compression. Place the cat on his right side, laying him on the firmest surface possible. Place your thumb on the side of his sternum (chest) that is facing up and wrap your other four fingers around the chest so that your fingers are underneath his body and pressing on the other side of the sternum. Squeeze the chest firmly between the thumb and four fingers. Release. Then squeeze firmly again. Release. Repeat 6 times, then wait 5 seconds to see if the chest expands and whether any pulse or heartbeat results. If it doesn't, repeat. When you combine pulmonary resuscitation with external heart massage, you should perform 1 pulmonary expansion for every 6 cardiac compressions.

PREVENTION: Remove small knickknacks and other swallowable items from your cat's reach.

Wheezing

RELATED SYMPTOM: The noise sounds typical of a human with severe asthma.

POSSIBLE CAUSE: Has your pet been stung by an insect, confronted with an airborne substance (such as a cleaning product or wall paint), or had an unusually stressful day? Wheezing is usually attributed to narrowing of the larger airways. **Bronchoconstriction,** as the condition is called, is a bit mysterious. Although it's often caused by a type of allergy or an insect sting, it often occurs for absolutely no known reason.

CARE: Whether you suspect an allergic reaction, a bite, or are completely clueless as to the cause of your pet's wheezing, immediately take your cat to the veterinarian. Should your vet determine your cat is suffering from bronchoconstriction, she will administer a bronchodilator, cortisone injection, and/or antihistamine.

If you are unable to get to the vet right away, you may want to administer the following to your cat: vitamin C, vitamin B_6, vitamin B_{12}, and an antihistamine such as chlorpheniramine (Chlor-Trimeton is a recommended brand) or Benadryl (see Appendix E: List of Recommended Dosages, pp. 169–189).

PREVENTION: In the case of allergy-and insect-caused ailments, keep your cat away from stinging insects and household chemical products such as room deodorizers and cleaning substances.

Coughing

Just like a human's cough, a cat's cough is marked by a sudden, noisy expulsion of air from the lungs. Its purpose is a protective one: Coughing is a normal reflex that removes undesired material from the respiratory tract—especially the trachea and large bronchi. This unwelcome matter can be an irritant, a piece of food, or phlegm.

Coughs fall into two general categories: nonproductive and productive. A dry, nonproductive cough often sounds harsh or hacking. With a moist, productive cough, the cat may stretch his neck forward and/or lower his head in order to expectorate mucous. You may not see this phlegm, however, because cats often swallow any secretions they've coughed up.

Feline coughing can sometimes be confused with other symptoms, specifically gagging, retching, vomiting, regurgitating, sneezing, wheezing, and **bronchial spasms** (also known as reverse sneezing). **Coughing** expels air forcefully through an open mouth; **gagging** is a throat spasm that prevents matter from traveling down the throat; **retching** is a stomach spasm that occurs before vomiting; **vomiting** is a reflex that forcefully expels the contents of the stomach; **regurgitating** is the involuntary, non-propelled return of undigested food to the mouth after swallowing; **sneezing** is a forceful expiratory event during which the mouth is usually closed; **wheezing** is a type of labored breathing, which sounds similar to the noise made by human asthma sufferers; and **reverse sneezing** is a repetitive, forceful, inspiratory, involuntary snorting noise usually made when a cat is trying to clear the back of his nasal passages.

Depending on what is causing the coughing, it may be accompanied by hoarseness, shortness of breath, discharge from the nose or eyes, blood, decreased appetite, or fever.

Dry Cough Accompanied by Breathing Difficulties

RELATED SYMPTOMS: The cough may produce blood and the cat may vomit. The following symptoms may also be present: fainting spells, rapid breathing, increased pulse rate, lethargy, weight loss, loss of stamina, and a swollen belly due to fluid buildup in the abdomen.

POSSIBLE CAUSE: Are mosquitoes prevalent in your locale? Does your cat spend time outside? Have you been lax in giving your cat heartworm-preventive medicine? If so, there's a chance he may have **heartworm.** The heartworm parasite is spread by the mosquito, which bites your cat, simultaneously passing its larvae into your pet's bloodstream. Once there, the larvae migrate to the heart, make a home for themselves, and grow into adult worms. In turn, these worms send their larvae into the bloodstream for another mosquito to suck up and deposit into yet another cat. The worms themselves are long, threadlike things (often 14 inches) with small mouths: If not treated, they can live lodged in your cat for 5 years, which is the average lifespan of a heartworm.

By the time the worms have lived in your cat for 6 months, they are fully mature. Unfortunately, the small size of the feline heart and blood vessels means one or two worms are enough to cause major damage or even death. An inflammatory response to the parasites may develop and your pet's pulmonary arteries may become enlarged. As these worms die, they can lodge themselves in a blood vessel, blocking it and stressing the heart and lungs to the point where some cats die of heart attacks or pneumonia.

CARE: If your cat is vomiting frequently, fainting, or coughing up blood, **take him to the vet immediately**. Keep the cat extremely quiet (i.e., cage rest is advisable, *with absolutely no exercise*).

To determine what ails the animal, your vet will give him a thorough exam to rule out other respiratory illness. This checkup will feature a blood test to detect both the presence of the heartworms' larvae and the specific proteins that the adult worms shed into the bloodstream. Blood testing is recommended, but usually not required, before heartworm medication is administered. Some vets may also choose to perform a chest X ray to look for the enlarged pulmonary arteries indicative of heartworm infestation. Cat owners should be aware that feline heartworm disease is harder to diagnose and treat successfully than its canine counterpart.

Ridding the body of worms requires an arsenical compound administered intravenously by your vet 2 times a day (A 2- to 7-day hospital stay is commonly recommended.) The goal of treatment is to kill the worms slowly so that the dead worms gradually pass into the lungs, where your pet's immune defenses eliminate them. Worms begin dying 5 to 10 days after therapy. During this time, strictly confine your pet indoors and keep exercise or excitement to a minimum, because both increase blood flow in the pulmonary arteries, upping the odds that a large mass of worms will be carried into the lungs. This can block the pulmonary blood flow and cause a severe reaction, including coughing, pneumonia, and fever. Such reactions must be treated immediately with antibiotics and corticosteroids.

Approximately 4 to 6 weeks after the adult worms have been treated, a second medication is administered to kill any larvae still swimming in the bloodstream.

PREVENTION: Because some cats with heartworm show no apparent signs, have your pet's blood tested every 6 to 12 months if your cat gets outside and/or you live in a mosquito-friendly, heartworm-prevalent area. Keep your cat inside during mosquito-active times, such as late afternoon and evenings. If your locale is only sparsely populated with mosquitoes, a once-yearly checkup is still a good idea. Antiheartworm drugs are available and highly recommended in mosquito-infested and heartworm-prevalent areas.

Fits of Dry Cough, Accompanied by Wheezing and Breathing Difficulty

RELATED SYMPTOMS: The cough will be deep and will have come on suddenly. The cat's breathing will be open-mouthed and strained. He may sit up while hunching his shoulders or lie on his chest. In more severe cases, his tongue and gums may be pale and have a bluish or grayish hue.

POSSIBLE CAUSE: Is your cat a Himalayan or Siamese? Has he been exposed to harmful, irritating airborne substances? Did he inhale foreign matter (sawdust, pollen, sand, dust, or feathers) days earlier that could possibly be irritating his windpipe or lungs? If so, it's possible that he has **feline bronchial asthma**, an acute respiratory disease that closely resembles its human counterpart. Feline bronchial asthma is characterized by spontaneous constriction of the bronchial passages. The coughing and wheezing are caused when the cat's airways begin to spasm when touched by an inhaled environmental allergen or irritant. It should be noted that feline bronchial asthma is usually sparked by an unknown cause.

CARE: Cats with acute onset of respiratory distress **should be taken to the vet immediately**. Because many ailments (from heart failure to cancer) can produce respiratory distress and coughing, X rays and an endoscopic examination are often performed. When coughing is continual, cough suppressants may be considered. Bronchodilators, supportive oxygen therapy, adrenaline, and/or cortisone may be immediately administered to relieve the bronchial spasms and allow the cat to breathe more freely. In severe cases, your cat may be sedated and hospitalized overnight.

If an offending allergen can be blamed, your vet will ask you to remove it from your cat's environment—an impossible task if your cat is allergic to the pollen thrown off by neighborhood plants. In such cases, or in instances where the cause remains unknown, be prepared to administer low doses of oral cortisone every other day, perhaps for the entirety of your cat's life.

Though bronchial asthma may be controlled, it cannot be cured. The cat's airways will always be prone to convulsing in the presence of specific allergens. Yet, just like asthmatic humans, most cats with feline bronchial asthma are otherwise healthy. You can keep your kitty more comfortable by keeping the house as dust-free as possible (electronic airfilters may prove helpful for this). Certain types of inside plants may need to be removed, whereas others may be helpful. You should avoid using room deodorizers, strong chemical cleaners, and fabric softeners, and try not to smoke around your pet.

PREVENTION: If you know a particular allergen sparks your cat's asthma, try to keep your cat away from it. Cats with feline bronchial asthma should be kept inside. Antioxidants such as sulfur and vitamins A and C are often helpful in treating respiratory allergies and in reducing the frequency and/or amount of cortisone therapy (see Appendix E: List of Recommended Dosages, pp. 169–189).

Persistent Dry Couch

RELATED SYMPTOMS: The cough may be mild or harsh, but it is persistent. Breathing difficulties may also be present in severe cases.

POSSIBLE CAUSES: Has your cat been outdoors, where he may have stepped in or licked the stool of another cat or dog? Could he have eaten a bird or some grass contaminated by feces? If so, he may be infected with one of several types of **parasitic worms**, such as **roundworms, hookworms, lung flukes**, or one of a number of other respiratory worms. Once inside the body, these worms live happily in the lung tissue and bronchial passages, where they reproduce. When the cat coughs, the parasite's eggs are brought up into his mouth, then swallowed. The eggs pass into the cat's stomach and intestines, then either exit the body through the anus or remain in the intestinal tract, where they develop into adult worms.

CARE: Respiratory worms—with the exception of heartworms (see the section, Dry Cough Accompanied by Breathing Difficulties, pp. 82–84)— are usually not a serious health threat, but do require a visit to your vet.

If you suspect your cat is infested with worms, take him to a vet because the worms can suppress his immune system. In order to identify the type of parasite present, your vet will look for microscopic worm eggs, then prescribe the appropriate worming medicine to rid your cat's body of the pests.

PREVENTION: Keeping your cat indoors will prevent him from coming into contact with the contaminated stool of other animals. If yours is an outdoor cat, take him to the vet at least every 6 months to have his feces checked for worms. Worm infestations should be taken seriously because they are also public health hazzards: Keep children away from areas you suspect may be contaminated, such as a sandbox that a cat may have used as his litter box.

Repeated Dry Coughing That Is Worse After Physical Activity

RELATED SYMPTOMS: If your cat has been coughing for 2 or more days, the cough may turn croupy-sounding, possibly accompanied by the gagging up of foamy saliva. At this point, he may also make periodic gagging movements and may seem to be swallowing repeatedly. He may hunch his shoulders and stretch his head and neck downward in an attempt to expectorate phlegm.

POSSIBLE CAUSE: Has your cat been exposed to fumes, pollens, dust or another respiratory irritant? Has he been diagnosed with or exposed to feline viral respiratory disease? There's a chance your cat may have **bronchitis,** a condition caused by inflammation of the bronchial passages. When the bronchial passages become inflamed, a cat attempts to cough out whatever is irritating his airways. When that doesn't work, the airways respond by producing a thick coating of mucous in an attempt to trap the invader with this viscous material. When the irritant persists, the coughing and mucous production continue and a secondary infection may develop.

CARE: Bronchitis frequently leads to pneumonia if left untreated. If allowed to become severe, bronchitis also may damage the bronchi. In the worst-case scenario, constant coughing can destroy the lungs and air sacs, and it can develop into an irreversible condition called emphysema. Therefore, you would be wise to take your pet to the vet at the first sign of illness. To reach a diagnosis, she will give your cat a thorough exam, carefully listening to his lungs with a stethoscope and examining the mucous. If the condition is severe, chronic, or persistent, a chest X ray may be performed. Your veterinarian may use antibiotics, cough suppressants, expectorants, and humidifiers to cure the disease.

At home, keep your cat warm and the air in your home humidified. For the first week, you may want to confine your kitty to a draft-free room in which you place a vaporizer. Once your cat has had bronchitis, there's a good chance that he may more susceptible to the condition at other points throughout his life. Therefore, you must take pains for the rest of your kitty's life to keep him out of dry, cold air, which irritates the respiratory tract. As an important supportive measure, give your cat antioxidants such as sulfur and vitamins A and C to strengthen his immune system (see Appendix E: List of Recommended Dosages, pp. 169–189).

PREVENTION: Keep your cat away from fumes and dust. Address all coughing immediately before it worsens.

Single Episode of Dry Coughing

RELATED SYMPTOMS: The hacking-sounding cough produces no mucous or blood and is not accompanied by fever or swollen glands.

POSSIBLE CAUSE: Has the cat recently been fed? Has he been rooting in dirt or underfoot during housecleaning? If the cough appears directly after any of these activities, he may have **inhaled a foreign object.** When a particle of food, liquid, plant material (such as a seed, branch, or bone), dirt, or other foreign material travels down the windpipe, it heads straight for the lungs. Coughing forces the object out of the airway and back into the throat, where it is spit out or makes its way down the correct passageway—the esophagus, which leads to the stomach. (A cough resulting from a foreign body in the windpipe may sound much the same as a gag resulting from a foreign body in the esophagus or back of the throat.)

CARE: Observe your cat. He should cough freely for 1 minute, then stop once the windpipe has been cleared. If the coughing continues and/ or is accompanied by a refusal to eat, restlessness, blood, and/or excess saliva, a piece of food or some other foreign object may have become lodged in the animal's throat or windpipe. Contact your veterinarian.

If coughing persists for more than 5 minutes without stopping and/or the cat appears to weaken, has a hard time breathing, and his tongue or gums look blue or grayish, consider performing the Heimlich maneuver. If the cat is conscious, lie him on his side—preferably on a table—and position your hands around his rib cage just behind his last rib. Give 4 quick, forceful, slightly upward thrusts. This movement forces the diaphragm up while producing an exhalation of air that is usually forceful enough to dislodge whatever is stuck. Peer inside his mouth to see whether the object has been dislodged. If not, repeat the thrusts. If the cat passes out, look down his throat for a foreign object that you can remove by yourself and then **rush him to the nearest vet.**

Should the cat lose consciousness, you may want, to perform CPR. Start by opening his mouth and lying the tongue to one side between the top and bottom molars. Place your finger in the cat's mouth and throat to feel for any obstructions, including vomit, mucous, or blood. Remove anything you find. *Extend the cat's head and neck, then close his mouth. Inhale deeply, completely cover your pet's nose and mouth with your mouth, then exhale into his nostrils.* The air should reach his chest: Watch for the chest to swell. Remove your mouth and allow the cat's chest to deflate normally. When it has, put your mouth over his nose and mouth and start again. This inflate-deflate cycle should be done 12 times per minute, until the cat begins breathing on his own.

Often, immediately after a cat has ceased breathing (or just prior to it), his heart may stop. (You can check for his heartbeat by wrapping your hand around his chest, just behind his front legs, and applying slight pressure. Alternately, you can check the cat's pulse, which is best felt on the inner side of the thigh in the groin area.) If the nearest vet is some distance away, you also will have to perform external cardiac compression. Place the cat on his right side, laying him on the firmest surface possible. Place your thumb on the side of his sternum (chest) that is facing up and wrap your other four fingers around the chest so that your fingers are underneath his body and pressing on the other side of the sternum. Squeeze the chest firmly between the thumb and four fingers. Release. Then squeeze firmly again. Release. Repeat 6 times, then wait 5 seconds to see if the chest expands and whether any pulse or heartbeat results. If it doesn't, repeat. When you combine pulmonary resuscitation with external heart massage, you should perform 1 pulmonary expansion for every 6 cardiac compressions.

Once at the vet's, she will provide supportive oxygen if necessary and determine the nature of the obstruction, performing a radiographic or endoscopic examination in an attempt to view the object. The object is usually extracted from the windpipe or throat with a forceps and or endoscope while the cat is under anesthesia. Should the object be lodged in the esophagus rather than the windpipe, your vet may simply push it into the stomach. If an infection has set in, the animal will be put on a soft-food diet for 1 to 2 weeks and given antibiotics.

PREVENTION: Never give your cat bones of any type to chew on. Furthermore, ban all small balls, sharp toys, string, dental floss, yarn, thread, and needles.

Wet Cough, Accompanied by Depression and Fever

RELATED SYMPTOMS: The cough will be weak and perhaps painful—and can be artificially triggered by tapping your pet's chest. In addition, your cat may have nasal and eye discharge, weight loss, lethargy, and labored breathing with shorter, more rapid breaths than usual.

POSSIBLE CAUSE: Has your cat recently had bronchitis or an upper-respiratory infection? Has he undergone surgery or been anesthetized for any reason (both of which can lead to accidental inhalation of material from the stomach into the lungs)? Has your pet been exposed to noxious fumes within the past 24 to 48 hours? Any of these can lead to **pneumonia**, an inflammatory condition of the lung. Any animal can get pneumonia, but it most often strikes those under two years of age and older than eight years.

CARE: To attempt to treat your cat's symptoms at home, provide him with lots of fluids and the antioxidants sulfur and vitamins A and C (see Appendix E: List of Recommended Dosages, pp. 169–189). Be sure he gets abundant rest in a warm room with 50 to 60 percent humidity. If 48 hours of homecare hasn't improved the symptoms, take your pet to the vet. Quite a few illnesses share symptoms with pneumonia, which is why your vet will take chest X rays to help reach a diagnosis. To determine the underlying cause of the pneumonia, samples of lung secretions are obtained for bacterial and fungal cultures. This is done using general anesthesia.

Treatment of pneumonia includes intravenous or subcutaneous (beneath the skin) fluid and antibiotic administration, maintenance of normal body temperature, rest, and perhaps expectorants. Coax your cat to gently walk around the house at least 3 times daily. This stimulates the cough reflex that can help break up and expectorate mucous. Depending on what is causing the pneumonia, specific antibiotics or antifungal drugs may be prescribed.

PREVENTION: Because many seemingly not-so-serious upper-respiratory ailments can worsen into pneumonia, the best prevention is to treat these conditions immediately before they can become life-threatening.

Pale or Blue Mucous Membranes

Most of us don't spend much time studying our pet's mucous membranes. Perhaps that's because this particular tissue—comprised of the gums, tongue, and the roof of the mouth—isn't immediately visible. It takes some work to pry open your cat's mouth and have a peek inside. Yet, attempting this feat regularly—say, 1 time a week or when you brush your cat's teeth—is worth the effort. Should you notice white, yellow, grayish, or bluish membranes, you can be sure your cat's not well.

Pale to white gums and oral membranes—as opposed to pink ones—are a result of **anemia** (see the following section, Shallow Breathing with Fast Heart Rate, Pale Tongue and White Gums, and Lethargy). Anemia develops when the concentration of hemoglobin or the number of red blood cells, which are responsible for carrying oxygen to the rest of the body, becomes diminished. If the anemia is severe, pale to white gums will result.

The symptom that describes blue or gray gums and mucous membranes is called **cyanosis,** but in order to recognize it, you first must know the normal color of the inside of your cat's mouth. The blue or gray hue is due to a lack of oxygen being delivered to the body's tissue—for example, if the cat's lungs or heart are diseased or the animal is choking. If the lack of oxygen is severe or prolonged, the skin in addition to the mucous membranes, can also appear bluish or grayish.

Lack of nourishing, oxygen-rich blood can be blamed on a number of ailments: a **heart condition, heartworms,** a **pulmonary embolism, pneumonia,** or a **disorder of the capillaries**. The specific cause of the symptom must be identified and treated before the symptom will disappear. Blue cyanotic and white anemic gums are both a result of poor oxygen flow to the tissue, but result from different causes. (Also see section in Chapter 3, Enlarged Lymph Nodes, Pale Mucous Membranes, Recurring Infections, Apathy, Loss of Appetite, and Fever, pp. 29–31.)

Shallow Breathing with Fast Heart Rate, Pale Tongue and White Gums, and Lethargy

RELATED SYMPTOMS: The breath is also faster than usual and the animal tires quickly when exercised. The cat is sleeping increasingly more and is reluctant to play even his favorite games. A fever of 103° to 106°F might be present, and your kitty may not be interested in food.

POSSIBLE CAUSE: Has your pet recently been diagnosed with heartworm disease? Is his coat infested with fleas and ticks? Has he been treated for some internal condition? Do you give your cat a primarily vegetarian diet? Did his mother harbor the *hemobartonella felis* organism, or could he have been bitten by a cat who was infected with it? Has he experienced a large loss of blood? If the answer to any of these is "yes," your cat might be anemic. **Anemia** typically develops over a period of time and is brought about by anything that lowers the body's red blood cell count—from diet to environmental toxins to prescription drugs. Because the red blood cells are what carry energy-giving oxygen to all of the body's tissues, treating the cat as soon as the condition is diagnosed is imperative.

CARE: If you suspect your cat is anemic, take him to the vet, who will perform a simple blood test, known as a hematocrit. This test will determine the presence and degree of anemia. Treatment of the condition depends on the cause: Ridding the body of parasites, changing the diet, or treating an illness may be enough to up the red blood cell count. In severe cases, blood transfusions and oxygen therapy might be necessary. Provide your cat with a superior quality, chemical-free, high-protein diet, as well as complete multi-vitamin and trace-mineral supplements (ask your vet for a specific recommendation).

PREVENTION: A weekly home exam will help you detect anemia in its early stages. A high-quality diet may prevent nutritional anemias.

Sneezing

Sneezing is not an illness. It's the body's protective response to an irritant in the nasal passages or sinuses. A feline sneeze is really no different from a human's. A forceful expulsion of air travels through the airways at great speed. In contrast to a cough, the mouth usually remains closed. In both person and cat, the sneeze is a reflex that helps clear the respiratory passages. Sneezing commonly accompanies an acute condition, rather than a chronic disorder.

Bouts of Sneezing with Fever

RELATED SYMPTOMS: Between 2 to 3 days after the sneezing and fever appear, the cat develops clear nasal and/or eye discharge that contains mucous. Eventually, this discharge may also contain pus. Your kitty may breathe with his mouth open. He may also refuse food and water, and he may drool.

POSSIBLE CAUSES: Did you recently adopt your kitten from a cattery, pound, or breeder? Was your cat recently in contact with another cat or another cat's litter box? Did you recently touch your own pet after playing with a sneezing cat? There's a chance your kitty has **feline viral respiratory-disease complex,** one of the most common infectious conditions encountered by cat owners—and vets.

The condition is actually caused by one of two virus families: the **herpes virus group,** which produces **feline viral rhinotracheitis;** and the **calicivirus group,** which is responsible for **feline calici viral disease.** These viruses are transmitted feline-to-feline by contact with infected saliva, nasal secretions, eye discharge, water and food bowls, litter boxes, and humans carrying the virus on their skin. Yet, no matter which of these two viral families is responsible for the infection, the signs are so similar that they often can only be told apart by special blood tests. Regardless of which viral group is at work, the actual clinical signs of illness usually appear 2 to 17 days after exposure, reaching their peak at 10 days postexposure.

CARE: Take your cat to the vet at the first available opportunity— immediately if he is a young kitten, elderly, or especially frail. In the meantime, if you own other cats, isolate your sick kitty to prevent cross-transmission. Don't forget to wash your hands after playing with your ailing kitty: you don't want to give the virus a chance to find more victims!

By observing clinical signs, performing a thorough physical exam, and running a few simple laboratory tests, your vet will be able to reach a diagnosis. If your cat does, indeed, have a viral infection, your veterinarian will place your cat on oral antibiotics to prevent any secondary bacterial infections while the body's immune system fights off the virus.

Back at home, remove secretions from your pet's eyes and nose by dabbing them with a soft cloth dipped in warm water or chamomile tea. Also, keep the room's humidity high, perhaps with a home vaporizer: This will keep secretions liquefied, making it easier for your pet to breathe. Because many ill animals are weak and not interested in eating, your vet may suggest bribing your cat with a highly palatable food such as strained meat baby food—a treat few cats can resist. Likewise, if your kitty isn't drinking, ask your vet for a syringe that you can use to shoot a small stream of water into the side of the little one's mouth.

If yours is a multicat household, expect to keep the sick fellow in quarantine until the vet gives you the okay to let him mingle with his housemates. Providing your cat with the antioxidants sulfur and vitamins A and C will help strengthen his immune system. A combination of the herbs Echinacea and goldenseal is helpful in treating both viral and bacterial functions (see Appendix E: List of Recommended Dosages, pp. 169–189).

PREVENTION: Keep your cat away from felines who exhibit signs of a respiratory disease. Wash your hands after playing with other cats. Have your cat vaccinated against these serious respiratory viruses. Place your cat on a high-quality diet and have him checked for intestinal parasites, which weaken the immune system.

Chronic Sneezing, Noisy Breathing, and Foul-Smelling, Watery Nasal Discharge

RELATED SYMPTOMS: The cat frequently paws at or rubs his face. Any combination of the following may also be present: bloody nasal discharge, reverse sneezing, pus and scabs on the sides of the nose, and plugged nostrils.

POSSIBLE CAUSE: Did your cat come from a cattery or humane society? Has your cat had a recurrent nasal discharge since he was a kitten? Has your cat suffered from some type of upper-respiratory condition in the past? It is possible that your cat has **rhinitis**—which in vet-speak means a **feline head cold** complete with inflamed nasal passages. This can be caused by a virus, a bacterial infection, or even an airborne allergen. In some cats, rhinitis can be chronic and recur whenever the cat is stressed.

CARE: If it becomes chronic, rhinitis can damage nasal cartilage. Be safe rather than sorry and take your pet to the vet. She will examine the nasal discharge and listen to your cat's breathing. If your vet thinks the rhinitis has progressed to bronchitis and pneumonia, she may want to take chest X rays. Depending on what is causing the ailment, she may recommend giving your cat antihistamines, antibiotics, corticosteriods, and/or cough suppressants.

In the event that you are unable to reach the vet for a few days, you can try giving your cat an oral antihistamine like chlorpheniramine (a recommended brand name is Chlor-Trimeton) to dry up his runny nose (see Appendix E: List of Recommended Dosages, pp. 169–189). Cleaning your kitty's nasal passages carefully with a Q-Tip to clear away any mucous can help him to breathe easier and make him more comfortable.

PREVENTION: Keep any nasal condition from getting worse by keeping your cat's nose clean. This is easily done with a soft cloth dipped in warm water or warm chamomile tea (the chamomile has a soothing effect). Megadoses of vitamin C and vitamin A along with a sulfur supplement will help the body deal with bacterial and viral infection and reduce allergic reactions. (see Appendix E: List of Recommended Dosages, pp. 169–189).

Reverse Sneezing (Repetitive or Spasmodic Snorting)

At some point you may have heard a forceful, spasmatic snorting noise burst from your pet's nose. Called a reverse sneeze, the sound is very dramatic—and frightening. Fortunately, reverse sneeze attacks are rarely anything to worry about. More good news: Reverse sneezing ordinarily occurs in cats who are otherwise healthy. It is usually brought on by something relatively harmless, such as: water drinking, excitement, pressure from a collar, or a postnasal drip. The sound you're hearing is your pet trying to clear the back of his nasal passages. This usually takes from 3 seconds to 3 minutes.

It is theorized that cats with sensitive throats—due to irritation caused by **an incomplete opening of the epiglottis** after swallowing—are those most likely to reverse sneeze. Treatment is rarely needed.

A reverse sneeze is a startling, "here one minute, gone the next" kind of thing and should not be confused with the snoring or squeaking that an animal with a partial large-airway obstruction may make when trying to breathe. Reverse sneezing is often mistaken for coughing or retching.

If your pet suffers from attacks of reverse sneezing several times a day or for longer than 30 seconds on a regular basis, talk to your veterinarian to be sure the noise is indeed a reverse sneeze. If the reverse sneezing occurs so frequently that it becomes annoying or upsetting to you, ask your vet for help. If an allergy is causing a postnasal drip, antihistamine therapy may be in order. Keep your cat away from dusty areas and any chemical sprays that may cause nasopharyngeal irritation. If you think that your cat may be allergic to tree, grass, or weed pollens, keep him inside.

CHAPTER 8

Abdomen

When vets talk about your cat's abdomen, they aren't just referring to the stomach. The abdomen is a large hollow cavity filled with various organs. The stomach is one of these organs, but also included are the kidneys, liver, bladder, intestines, and reproductive organs. A problem with any of these can manifest itself as a tummyache or with bloating, bleeding, a change in elimination habits, lethargy, bad breath, and a variety of other subtle-to-showy signals. A problem with the abdominal wall itself, like a tear in its tissue or a hernia, also may show a variety of seemingly unrelated symptoms.

Because illnesses involving the internal organs can be serious—or even fatal—keep an eye out for any behavioral changes (or even a barely perceptible deviation) in your pet.

Bloated, Distended, or Painful Abdomen

A belly can become swollen for any number of reasons. There are the obvious ones. Your cat has just eaten a big meal, eaten rotten food (maybe she's gotten into the garbage), or drunk a large volume of water.

Other causes may be harder to pinpoint because they produce a slow, gradual bloating, perhaps over 1 day's or 1 week's time. Conditions that commonly cause gradual abdominal distension include **constipation, parasites, kidney failure, liver disease, heart disease,** a **hormonal imbalance, abdominal fluid, abdominal tumors, pregnancy** (or **false pregnancy**), a **hernia, obesity,** or **abnormal organ enlargements.**

Because a large number of different illnesses cause bloating, your vet will narrow the diagnostic field by viewing a swollen abdomen in light of all the cat's accompanying symptoms, such as vomiting, coughing, restlessness, a change in elimination habits, or increased drinking.

As for abdominal pain, you'll know your cat's abdomen hurts if she shrinks from being touched there, moves cautiously or not at all, and adopts either an arched-back stance or what vets call a prayer pose (back legs standing, front legs outstretched and lowered onto the ground, head resting on front legs). A cat with a stomachache might refuse food, tremble, hide, and/or cry. A tender tummy most commonly indicates an injury or disease of one of the organs housed in the abdomen.

Obvious Bulge at Midabdomen, Groin, or Rectal Area

RELATED SYMPTOM: The cat may appear restless.

POSSIBLE CAUSES: Has your cat been in an accident lately? Is she a young kitten? Has she recently overexerted herself physically? Is she a pregnant female or a new mother? Your pet may have one of several types of hernias, including an **umbilical hernia** (at the navel), an **inguinal hernia** (of the groin area), or a **perineal hernia** (alongside the anus). Standing the animal on her hind legs helps you to see an inguinal or perineal hernia as it bulges out. Umbilical hernias can be easily seen when the cat is standing normally.

Hernias happen when interior muscles tear after being suddenly strained beyond their capacity. These muscles act as a kind of containing wall to hold organs in place. A rip in this muscular wall allows fat or organs to "poke through" the injury and situate themselves directly beneath the skin: These runaway organs comprise the hernia's visible surface bulge. Depending on the type and extent of the hernia, the cat may feel little discomfort or a great deal of pain. Hernias can be life-threatening if a loop of intestine passes into the hernia and becomes twisted, losing its blood supply.

CARE: Visit the vet, who can easily diagnose the problem with a physical exam. If your cat does have a hernia, your vet may want to perform an operation to repair it.

PREVENTION: Although hernias cannot be prevented, the sooner the problem is addressed, the less discomfort and health risks your kitty suffers.

Change in Appetite and/or Weight

No one wants to read a lecture about the perils of letting a cat get fat. Yet, just like it is in human health, obesity is a primary factor in many types of feline illness. A kitty weighing over 15 percent more than the standard accepted weight for her bone structure is considered over-weight. Just how serious is this extra body fat? A portly kitty is at in-creased risk for **musculoskeletal, cardiovascular, gastrointestinal, endocrine, respiratory, immune,** and **reproductive disorders,** including **cancer.**

Common, nonmedical factors contributing to the feline world's grow-ing girth are predictable ones: high-fat foods (many of the so-called "pre-mium" brands of cat food are high in fat), lack of adequate exercise (running to the food bowl at dinner time isn't enough), and being fed little cat treats throughout the day. Correct one or more of these situa-tions and an overweight kitty will usually slim down.

There are instances, however, when obesity can only be blamed on a medical condition that increases a cat's appetite, putting her in a con-tinual state of ravenousness. These illnesses most typically involve a hor-monal imbalance. The affected cat eats dinner, continues to act hungry, and is given more food by her attentive owner. Unfortunately, after a few months of this, the cat begins to get pudgy.

But what if you monitor your pet's diet, wouldn't dream of feeding her table scraps, give her plenty of exercise, and notice her growing fat anyway? You'd probably worry that something was wrong—and rightly so. Changes in food intake or weight are good indicators that your cat might be ailing. In the foregoing example, a slower metabolism due to age, illness, or an increasingly sedentary lifestyle is a strong possibility: A cat who isn't burning calories is going to store them as fat.

What about the kitty whose diet and lifestyle haven't changed (mean-ing she's not consuming less calories or burning more) yet is still losing weight? The only explanation is that the cat isn't assimilating the calories and nutrients she is taking in, a common occurrence with **diabetes.** You may even notice the cat becoming weak, lethargic, or apathetic. Diseases of the intestinal lining can result in loss of protein and subsequent weight loss. **Digestive-enzyme deficiencies** and **malabsorption syndrome** are ad-ditional causes for weight loss.

Of course, the most obvious road to weight loss is via appetite loss, which vets call **anorexia.** Unlike the psychologically based human illness of the same name, the condition in cats is a physical one that accom-panies a wide range of medical conditions—from internal infection or parasites, to cancer or other diseases of one of the abdominal organs, such as the liver. Simply put, an animal who doesn't feel well often can't stomach the thought of food—and without her regular calorie supply, she loses weight.

Increased Appetite, Vomiting, Increased Heart Rate, and Enlarged Thyroid Glands in the Neck

RELATED SYMPTOMS: Despite the increased food intake, your cat may progressively lose weight and have an increased volume of feces. Your cat may appear restless and irritable, and may pace. (You can tell whether your cat's thyroid glands are swollen by putting your thumb on one side of her windpipe and your forefinger on the other side. Hold your fingers there, then slide them down the cat's neck. If her glands are swollen, you'll feel a little bump.)

POSSIBLE CAUSE: Is yours an older cat? Is he underweight or losing weight despite having a large appetite? He may have **hyperthyroidism**, one of the most common hormone conditions among older and/or underweight cats. It occurs when the thyroid gland kicks into overdrive, producing an overabundance of circulating thyroid hormone. Because the excess hormone increases the animal's metabolic rate, the entire body is stressed.

CARE: Take your cat to the vet, who will perform a blood test and possibly a urinalysis. Should your cat be diagnosed with hyperthyroidism, your vet will try to regulate the animal's thyroid hormone production with antithyroid drugs. In severe cases, surgery may be needed to remove the thyroid gland.

To keep your cat's nutrition at an optimal level while keeping his weight at an acceptable level, your vet may recommend a special high-nutrient, easy-to-digest diet. Plant-derived digestive enzyme supplements are also recommended (see Appendix E: List of Recommended Dosages, pp. 169–189).

PREVENTION: There is no prevention.

Lack of Appetite, Chronic Vomiting, and Lethargy

RELATED SYMPTOMS: You might notice bits of curdled blood in the cat's vomit. The animal may have a painful abdomen and thus shrink from being touched there. Her feces may be dark or even black, indicating the presence of blood.

POSSIBLE CAUSES: Has your cat been under stress? Has she swallowed a sharp object or ingested a caustic substance? Does she have a metabolic or infectious disease? Has she been diagnosed with internal parasites? A "yes" to any of these seemingly unrelated questions can signal the presence of an **ulcer** or a **stomach tumor**.

Rather than being a condition in themselves, ulcers are actually a sign of another underlying condition. They are produced by an overabundance of gastric acid that erodes the gastrointestinal lining. A bacteria called *Helicobacter pylori* has been recently discovered to be linked to many ulcers. Scientists believe this bacteria may be a major cause of certain types of ulcers. What causes tumors, however, isn't known, although it's suspected that chronic stomach inflammation, often increased by ulcers, contributes to their growth.

CARE: Your vet will perform a radiograph or endoscopic exam to determine whether ulcers or tumors are present. For ulcers, the underlying cause will be treated. If the *Helicobacter pylori* bacteria is present, an antibiotic may be prescribed. The cat will also receive medication to reduce the stomach acid, thus giving the ulcerated areas a chance to heal. If tumors are present, your vet may operate to remove them.

PREVENTION: If administered on a permanent basis, drugs that check the production of gastric acid can prevent ulcers. Feed your pet the same amount of food that she usually gets, but divide it into smaller meals given 3 or 4 times a day. Feeding her a low-fat variety also may be helpful.

Lack of Appetite, Weight Loss, and Low-Grade Fever

RELATED SYMPTOMS: Your cat may have breathing difficulties and a swollen abdomen, or she may have a combination of coughing, vomiting, diarrhea, seizures, or blindness.

POSSIBLE CAUSE: Is your cat between three months and three years old? Did your cat come from a cattery or multicat facility? Does she spend time outdoors where she can come into contact with other cats? She may have **feline infectious peritonitis** (also called **FIP**), a contagious viral disease that is spread by inhalation or ingestion of body fluids. Most cats infected with FIP, as vets call the disease, either show no symptoms or show such mild ones that they are easily overlooked. Symptoms—if they do appear—typically show up to 3 weeks after being exposed to the virus. In fact, signs can appear for a few days, then disappear, only to reemerge 3 or 4 years later.

What makes this viral disease unique is that the actual organ damage that results from being infected with FIP is not actually caused by the virus itself, but by the immune system's response to the virus. To further confuse matters, there are two types of FIP: wet and dry. Your cat can have either or both. In the wet version, your cat's immune system attacks the blood vessels in response to FIP. Fluid oozes from the damaged vessels and accumulates within the chest and/or abdomen, causing breathing difficulties and/or a swollen abdomen.

With the dry version, numerous small nodules and localized areas of inflammation appear in various body parts, such as the lungs and heart, the gastrointestinal tract, the brain and spinal cord, the kidneys, and/or the eyes. Symptoms depend on the organs affected, but clinical signs can include coughing, vomiting, diarrhea, seizures, and blindness.

CARE: If you do suspect FIP, take your cat to the vet. A physical exam, blood test, and microscopic examination of chest and abdomen fluids can help to yield a diagnosis. There is no single laboratory test that is diagnostic for FIP and because of the variability in the symptoms, it may be very difficult to diagnose.

Unfortunately, no cure for FIP exists. Regulating your cat's immune response with steroids and chemotherapy drugs can temporarily relieve your cat's symptoms, but won't eliminate the virus. Your vet may recommend feeding your cat a diet that will entice her to eat, in order to curb the appetite loss and weight loss associated with the virus. Felines with clinical signs of FIP can survive anywhere from days to years depending on the degree of organ involvement.

PREVENTION: Be sure to keep a cat infected with FIP from coming into contact with other healthy cats. A FIP vaccine is available.

Large Appetite, Weight Loss, and Loose and Unformed Rancid-Smelling Stools

RELATED SYMPTOMS: The hair around the anus appears oily and/or greasy.

POSSIBLE CAUSE: Has your cat been diagnosed with a pancreatic or liver disease, lymphosarcoma in the intestine, or inflammatory bowel disease? A "yes" may indicate **malabsorption syndrome** or **maldigestion syndrome**. The condition occurs when something—a lack of digestive enzymes, in the case of liver or pancreatic disease; an injured bowel, in the case of lymphosarcoma and inflammatory bowel disease—keeps the small intestine from digesting food and/or absorbing its nutrients. The result is a malnourished cat with smelly, fat-laden stools.

CARE: Take your cat to the vet, who will perform an intestinal biopsy and test for the presence of sufficient digestive enzymes. Should your cat have malabsorption syndrome, your vet will concentrate on treating the illness that led to the symptom's development. The vet also will place your kitty on a low-fat diet and provide B-complex and fat-soluble vitamins to make up for the nutrients that your pet's body is excreting. Plant-derived enzyme supplements will aid in more complete digestion (see Appendix E: List of Recommended Dosages, pp. 169–189).

PREVENTION: Treat pancreatic and liver conditions, lymphosarcoma, and inflammatory bowel disease promptly. Feed your cat a highly digestible diet that is low in fat.

Weight Loss in Spite of Voracious Appetite, Accompanied by Excessive Thirst and Urination, Listlessness, and Lethargy

RELATED SYMPTOMS: The coat may appear dull and the cat may scratch it occasionally. The animal may develop a potbelly and bad breath that smells faintly of acetone.

POSSIBLE CAUSE: Is yours an older animal? Is she overweight? Is she a midsized cat? She may have **diabetes**, an illness (just like the human version) that originates with inadequate insulin production. The pancreas is responsible for manufacturing insulin, a substance that regulates the passage of sugar from the blood into the body cells. Without sufficient insulin, blood sugar rises, but the body cell sugar becomes depleted. This imbalance leads to kidney, liver, eye, and heart failure, and, eventually, death.

CARE: Determining whether your cat has excess sugar in her urine is possible with an at-home urine test available at your pharmacy (you can buy the same test used for humans). Keep in mind that normal urine does not normally contain sugar. Whether you try an at-home test or not (and regardless of the outcome), you will still need to visit the vet for a formal diagnosis, which will hinge on a simple blood sample. Should your kitty have diabetes, she will be placed on insulin therapy.

At the beginning of therapy, the cat usually stays under close veterinary supervision for the first 2 or 3 days while the proper insulin dosage is determined. From then on, however, you will be responsible for giving your pet daily injections, feeding her a strict low-fat, high-fiber diet doled out in 2 to 4 small meals per day, and providing round-the-clock access to drinking water. Because some diabetic cats have unexplained, spontaneous, once-in-a-while remissions, your vet may ask you to test your kitty's urine weekly (using the at-home strip method). Should you discover the sugar in her urine has decreased or increased, the necessary insulin adjustments can be made.

You can also supplement your cat's diet with zinc, chromium, and a standard antioxidant vitamin-mineral supplement. A high-quality digestive-enzyme supplement that is high in lipase should also be given to your pet regularly (see Appendix E: List of Recommended Dosages, pp. 169–189).

PREVENTION: Feed your cat a relatively low-fat, high-fiber diet that is free of simple sugars and chemicals. Don't let her become overweight.

Change in Thirst and Irregular Urination

In order to keep dehydration at bay, provide a constant supply of fresh water for your cat. But there may come a time when you notice your cat dipping into the water bowl more and more frequently—or perhaps infrequently. When not accompanied by other symptoms, a 1-or 2-day change in drinking habits is usually not cause for alarm. Maybe the weather is hot or your pet has depleted her body's water levels with more-vigorous-than-usual play.

If the abnormal thirst pattern continues, she could have a problem. **Kidney** and **liver conditions** generate increased thirst, as do many **internal infections** and **endocrine-system diseases** (see Chapter 6, Hair and Skin). A decrease in thirst is a less common sign of illness and usually is connected to pain or nausea: Your pet isn't drinking because it hurts to swallow (see Chapter 5, Mouth and Throat), or because she has learned that drinking results in nausea and vomiting.

On the topic of urination, a change in the amount of liquid your pet drinks is quite often accompanied by a change in her elimination habits. To know whether your cat is urinating abnormally, however, you first must know what is "normal" for your particular animal. Admittedly this can be tough if yours is an outside kitty; easier if your cat "does her business" in the litter box. Either way, it's important to keep abreast of your cat's elimination activity.

When discussing urination, the word "irregular" can refer to a change in the color of the urine or the frequency of elimination habits, obvious pain while voiding, or the presence of blood, mucous, or grit in the urine. Often one of these irregularities will be the only sign that your cat's urinary tract is not functioning properly. Other times, your kitty may also exhibit more obvious signs, such as vomiting or cowering when her tummy is touched.

Cloudy, Bloody Urine, Accompanied by Straining and Pain When Urinating

RELATED SYMPTOMS: Your cat may continually attempt to urinate while producing only a few drops of urine each time. The urine may contain blood clots or fine crystalline "urinary sand." The urine color may vary from dark yellow to pink to extremely bloody. Your cat may continually lick his or her genitals.

POSSIBLE CAUSE: Do you feed your cat a nonacidifying cat food? Do you sometimes forget to supply your cat with fresh water several times a day? Do you infrequently or haphazardly clean your cat's litter box? Your cat probably has **cystitis**, meaning an inflammation of the bladder. Bladders become inflamed for many reasons, but the most common cause is urolithiasis, urinary sand that results from mineral crystals precipitating out of the urine and forming a gritlike material. This grit or sand irritates the bladder wall and causes hemorrhaging and straining. Cystitis due to the formation of urinary sand can occur in both male and female cats, but is much more serious when it occurs in males because the crystals often produce a dangerous obstruction in the male cat's narrow urethra. (For further discussion on the condition in males, see Straining to Urinate but Producing No Urine, Crying, Attempting to Urinate in Places Other Than Litter Box, and Licking Genitals (in Male Cats), pp. 107–108)

Less common causes of cystitis are bacterial infections, tumors, and the formation of urinary stones, which can become quite large and may have to be removed surgically.

CARE: If you see your cat straining to urinate, consider the possibility that this is a potentially serious problem—especially if you have a male cat. Observe your cat closely while he is urinating and note the color, appearance, and quantity of urine he produces. Try to gather a sample of urine by placing strips of wax paper into a litter box that has been thoroughly cleaned with all the litter removed. When the cat urinates into this litter box, collect the urine and take it and your cat to the vet as soon as possible. In the meantime, provide your cat with fresh water to encourage drinking, which helps flush the bladder of crystals and any bacteria that may be present. Increasing your cat's water consumption also helps create a more dilute urine, which tends to produce less of a burning sensation during urination. Voiding frequently also allows less time for urine to become alkaline and, therefore, for further "grit" to form.

To determine whether your cat has cystitis and what is causing it, your veterinarian will examine your cat and palpate his or her abdomen, especially the bladder. He will either examine the urine you bring in, squeeze your kitty's bladder, or aspirate the urine from the bladder with a syringe and needle in order to obtain a sample. He will test the urine for the presence of blood, sugar, ketone, bile, and other substances and will examine the urine sediment for crystals, red blood cells, white blood cells, urinary casts, and epithelial cells. Furthermore, he will check the acidity and concentration of the urine. He may decide to take radiographs if he suspects a tumor or bladder stone.

Depending on what your vet determines the cause of the cystitis to be, he may recommend a special acidifying diet to help stop urinary sand formation or he may dispense urinary acidifying tablets (250 mg of vitamin C tablets given 1 to 3 times daily for a 8-to 12-pound cat can also be used). On the other hand, for infectious cystitis, he may inject and/or dispense antibiotics, which will be used for several weeks. Surgery may be necessary if either a tumor or large stone is causing the cystitis. A follow-up urinalysis is usually necessary to monitor the progress of the urinary problem.

PREVENTION: Provide your cat with plenty of fresh water. Clean your cat's litter box daily and change the entire box every 3 to 7 days in order to make her litter box as welcoming as possible. Some cats are very fastidious and hesitate to urinate in a somewhat dirty litter box. Feed your cat a high-quality, acidifying diet. Test your cat's urine periodically with special test strips (available from your vet or drugstore) to identify, with the help of your veterinarian, when the acidity of the urine is at its best level. Vitamin C in the form of ascorbic acid, when given daily in moderate doses (see Appendix E, List of Recommended Dosages, pp. 169–189), can help maintain the urine's proper acidity, which, in turn helps prevent crystal formation and precipitation into urinary sand. Don't take your cat's daily elimination behavior for granted; observe it as often as possible.

Increased Thirst and Urination, Vomiting, Weight Loss, Lethargy, and Dehydration

RELATED SYMPTOMS: The animal is apathetic, her breath smells like urine, and she seems either dazed or in a state of heightened excitability. The vomit may contain blood, and the stool is often dark and soft. The urine may be very light in color and very dilute.

POSSIBLE CAUSE: Is yours an older animal? Has she ingested poison at some point in the past? At some point in her life, was she on a long course (a month or more) of antibiotics? Has she ever been diagnosed with heat stroke, heart disease, repeat urinary-tract infections, periodontal disease, or an autoimmune disease? A "yes" to any one of these can indicate **chronic kidney failure** (as opposed to acute kidney failure).

Unlike acute kidney failure, chronic kidney failure is a slowly developing impairment that hinders the kidneys' ability to filter toxins from the blood. As kidney cells die—whether naturally due to aging or through chemicals or undue strain—they are replaced by scar tissue, which is useless at cleaning blood.

CARE: Take your cat to the vet, who will conduct blood and/or urine tests to diagnose the condition. The disease must be distinguished from a number of other diseases that cause excessive drinking and urination, particularly diabetes. Serious cases will be treated with intravenous fluids and a special low-protein diet. Antibiotics will be used if infection is present.

A cat with kidney disease should be kept as stress-free as possible in order to keep the body, and thus the kidneys, from having to work harder than necessary. A constant supply of fresh water should be available to the kitty, since even slight dehydration can make a bad situation worse. Because your cat will be urinating more, water-soluble vitamin B-complex may be prescribed to replace B-vitamins lost in the increased urine flow. Anabolic steroids may be prescribed to reduce muscle breakdown. Be sure to feed your cat a high-quality, chemical-free diet that is not acidic. If the cat is anemic, your vet may prescribe medication to stimulate red blood cell production. The antioxidants beta-carotene, vitamin A, and vitamin C (in the form of calcium ascorbate, not ascorbic acid) are recommended (see Appendix E: List of Recommended Dosages, pp. 169–189).

PREVENTION: Address any illness immediately and supply adequate fresh water at all times. Pay attention to your pet's drinking and elimination habits in order to detect changes early. Reduction of toxins in the pet's environment and in her food may slow progressive kidney damage.

Increased Thirst, Greatly Reduced Urine Output, Dark and/or Bloody Urine, and a Swollen Abdomen

RELATED SYMPTOMS: The cat may not be urinating at all. Her abdomen and loin area hurt and the animal may be feverish, apathetic, and refusing to eat. She may vomit and have diarrhea, which may or may not contain blood. Her breath may be sour and smell like urine. There may also be ulcers in her mouth and on her tongue.

POSSIBLE CAUSE: Has your cat recently been in an accident in which she lost a lot of blood? Could she have ingested rat poison, antifreeze, or any other toxic substance during an unsupervised moment? Does she suffer from stones in, or an injury to, the urinary-tract system? Has she recently wrestled with a bout of heavy diarrhea and vomiting? Has she been diagnosed with a serious abdominal infection, such as peritonitis or pyometra? Has she recently suffered from heat stroke? A nod to any of these may indicate **acute kidney failure.** The condition occurs when kidney cells are killed off due to chemicals, bacteria, trauma, or extreme strain caused from overuse, impairing the organ's ability to filter toxins from the blood.

CARE: **Take your cat to the vet immediately.** A blood and/or urine test can detect kidney failure. If her kidneys are failing, your pet will be given intravenous fluids and medications to stimulate kidney function. Your vet will then determine the underlying cause of the kidney failure and treat it.

PREVENTION: Supervise your cat's elimination behavior and address all symptoms promptly.

Straining to Urinate but Producing No Urine, Crying, Attempting to Urinate in Places Other Than Litter Box, and Licking Genitals (in Male Cats)

RELATED SYMPTOMS: Your male cat acts very restless and uncomfortable and may appear agitated or distressed. His meow may sound like a moan, and he may repeatedly attempt to urinate while standing or squatting and straining. He may make numerous trips to the litter box. If urine is produced at all, it may be blood-tinged or very red. If you look carefully, you may see fine crystals in the urine. If the problem persists for more than 24 hours, your cat may start to vomit and become weak.

POSSIBLE CAUSE: Is your cat a young male? Is he relatively inactive and overweight? Is he fed a mediocre diet that has not been specifically formulated to acidify the urine? Do you clean his litter haphazardly? If so, your male cat probably has **cystitis complicated by a urethral obstruction,** more commonly known as **blocked cat syndrome** or **plugged penis syndrome.** The condition occurs when a male cat's urine becomes concentrated with crystals, which then precipitate out of the urine and form grit that irritates the bladder wall and causes it to become inflamed and hemorrhage. The crystals, blood, and mucous produced in the bladder form a mass that moves down the progressively narrowing urinary tract, where it finally lodges and obstructs urine from passing out of the penis. This scenario is part of the larger picture known as FUS (Feline Urologic Syndrome). Although female cats can get FUS, they rarely obstruct because of their larger urethra and consequently are at far less risk of a potentially fatal outcome.

CARE: If the urethral obstruction is not relieved, the urinary bladder is not able to empty and progressively fills with urine. As it becomes greatly distended, it becomes more and more painful. Furthermore, the distended bladder puts fluid pressure on the kidney, which may injure the kidney and reduce its ability to remove toxic waste from the blood, and the cat may develop uremic poisoning. If the condition is left unattended long enough, the bladder could rupture. Because urethral obstruction is potentially life-threatening, **get your pet to your vet or the local veterinary emergency service as soon as you suspect this condition.** Your vet can quickly diagnose the problem by palpating the cat's abdomen, looking for a distended, painful abdomen. She may look at your cat's penis to see if the tip has been traumatized by the cat's rough tongue persistently licking the area. Upon milking and massaging the penis, your vet may be able to remove the offending plug if it has lodged very close to the tip of the penis. If that doesn't work, she will need to sedate your cat and insert a catheter into the urethra to flush the obstruction back into the bladder. Next, she will flush the bladder with a cold saline solution in an attempt to remove as much crystalline and mucous material as possible. The catheter may or may not be left in for 24 to 48 hours.

By the time the problem is discovered, the kidneys of blocked cats often have stopped removing toxins from the blood. These toxins build up and show elevated levels when your vet tests the cat's BUN (blood urea nitrogen). In such cases, the vet may put your cat on intravenous fluids to stimulate urination, reduce dehydration, and help flush toxins that have built up in the blood. If the bladder is severely inflamed and hemorrhagic, antibiotics may be given to prevent secondary bacterial infections. Short-term use of prednisone to reduce the inflammation in the bladder and urethra is a common approach.

PREVENTION: Feeding your cat a high-quality, acidifying diet or using a urinary acidifier such as vitamin C (ascorbic acid) may help maintain the urine's acidity below a pH level of 6.5. Encourage drinking by offering fresh water, which maintains the urine's acidic level by flushing the urine out of the bladder so it does not stagnate and become alkaline. Furthermore, this "flushing" reduces the crystal load in the bladder, which also minimizes the likelihood of a urethral obstruction. Most veterinarians can suggest a "prescription acidifying diet" specifically designed to reduce urine crystal formation. If your cat continues to obstruct despite the above suggestions, your vet may suggest a perineal urethrostomy, the operation that removes the penis, which is the area where the urethra narrows the most, thus enlarging the urethral opening to make it less prone to obstruction. Frequently cleaned litter boxes encourage more frequent urination, consequently eliminating the number of crystals in the bladder and lowering its acidity. Pay close attention to your cat's habits and see your vet when you first notice the above symptoms.

Flatulence

Once in a while, your cat may develop gas. If this happens infrequently, something in the cat's diet may be at fault. Has your kitty been allowed to wolf down a large amount of food in a small amount of time? Has she dined on table scraps of any kind? Smaller, more frequent feedings should remedy the first situation, while limiting (or banning outright) table scraps should treat the second. The only way to prevent the possibility that your cat is getting into the garbage or neighborhood pets' dinners during her travels is to supervise all outdoor play.

But what if your cat passes a more steady stream of gas? Before rushing to your vet, look at your cat's diet again. If you've introduced a new food, you might consider this as the reason. Put your kitty on her old food (or, if you switched brands for a specific reason, try yet another brand of food) for 1 week and see if the gas subsides. If this doesn't work, or if you haven't recently changed your cat's diet, he may have a gastrointestinal problem. Supplementing your kitty's diet with plant-derived digestive enzymes or *Lactobacillus* tablets can often solve the problem, and activated charcoal (available at your local drugstore) can absorb gas, toxins, or any other material irritating your pet's intestines (see Appendix E: List of Recommended Dosages, pp. 169–189). To help your vet diagnose a possible ailment, look for and list other symptoms before your office visit, such as a change in stool size or consistency, a change in defecation habits, vomiting, hunched posture, or weight loss.

Irregular Defecation and Feces

To know whether or not your pet is defecating normally, you first must know what is "normal" for your particular cat. Admittedly, this can be tough if yours is an outside kitty; somewhat easier if your cat "does her business" in an indoor litter box.

As unpleasant as the task seems, it's important to stay informed of your cat's defecation activity. Any problems here are reliable indicators of problems within the body—typically a **bacterial infection, a lodged foreign object, parasites,** or an **endocrine, gastric,** or **metabolic disorder.** Worth noting are: changes in the color, odor, and/or consistency of feces; changes in the frequency of defecation; the occurrence of obvious pain while defecating; and the presence of blood or mucous in waste materials.

Often one of these irregularities will be the only signal that your cat's health is off. Or your cat also may exhibit other more obvious signs, such as bloating, abdominal pain, flatulence, or vomiting.

Chronic Diarrhea and Weight Loss

RELATED SYMPTOMS: The diarrhea often contains mucous, and it may appear in cycles (perhaps occurring consistently for 1 to 2 weeks, then stopping, only to recur). Vomiting may also occur cyclically.

POSSIBLE CAUSES: Could your cat have internal parasites? Does your cat have food allergies? If you answered "yes" to either of these questions, your cat may have **inflammatory bowel disease (IBD)**. The cause of IBD, as the condition is also called, is unknown. Yet the illness is thought to be caused by bacteria or food proteins that irritate the intestine and activate the immune system. In response, the cat's body may be manufacturing antibodies that actually attack the cells of its own intestinal tract.

IBD actually describes three separate, yet related, illnesses that share identical symptoms: **eosinophilic enterocolitis, lymphocytic-plasmacytic enterocolitis,** and **granulomatous enteritis.** Eosinophilic enterocolitis affects the stomach, small intestine, or the colon. Although its cause is not exactly known, it's often seen in cats with internal parasites, such as roundworm or hookworm. Lymphocytic-plasmacytic enterocolitis is the most common type of IBD found in felines, and is associated with food allergies and and bacterial overgrowth in the large intestine. Vomiting sometimes accompanies it. Granulomatous enteritis is quite rare in cats and is similar to Crohn's disease, which affects humans. Simply put, this last type of IBD occurs when the fat and lymph nodes that surround the bowel become inflamed—through an unknown mechanism—thickening the bowel itself and narrowing the bowel's passageway.

CARE: Take your cat to the vet, who will eliminate other possible causes through an examination and lab testing. An endoscopic biopsy of the stomach, intestines, and colon will confirm the tentative diagnosis.

Should IBD be the culprit, your vet will begin by determining which specific type of IBD is affecting your cat. If it is eosinophilic enterocolitis, the vet may suggest a hyperallergenic diet that is low in fat and easy to digest. If lymphocytic-plasmacytic enterocolitis is behind the IBD, your vet will probably prescribe an antibiotic to treat the excess bacteria in the intestine or, in severe cases, prescribe an immunosuppressive drug. In the rare instance that granulomatous enteritis is to blame, your vet will place your cat on corticosteroids and immunosuppressive drugs, which can reduce inflammation and scarring. If the bowel passage has become extremely narrow, surgery may be necessary to widen it.

PREVENTION: Feed your kitty a high-quality, chemical-free, hypoallergenic diet accompanied by a flaxseed oil supplement and these antioxidants: vitamins A, C, E, and beta-carotene (see Appendix E: List of Recommended Dosages, pp. 169–189).

Diarrhea, Often Accompanied by Slimy Mucous and Bright Red Blood

RELATED SYMPTOMS: The diarrhea may or may not alternate with constipation. The cat may crouch forward and strain when defecating. The stool may have a foul smell.

POSSIBLE CAUSE: Is your cat fed "people food" or does she have access to the garbage can? Does she have internal parasites, such as coccidia or hookworms? Does she have food sensitivities? Has she been diagnosed with feline leukemia, feline AIDS, an immune disorder, a fungal infection, internal polyps, or tumors? A "yes" to any of these queries may indicate **colitis.** Colitis is an inflammation of the colon's mucous lining, which often results in painful diarrhea. The condition can be acute (a sudden, short bout) or chronic (long-term).

CARE: If the condition appeared suddenly in response to a change in diet, try withholding food from your cat for 24 hours. Slowly return to feeding your cat her original diet, or gradually, over a 2-week period, introduce a new diet of bland, hypoallergenic food—strained-meat baby food (made without stomach-upsetting onions) is ideal. Vitamin A has been found to help certain types of colitis. To soothe your pet's colon lining, try giving her roasted carob powder (found at health-food stores) 3 times a day for 3 days. You also can treat your cat's diarrhea with activated charcoal (available at your local pharmacy), which helps absorb toxins, poisons, and other irritating material (see Appendix E: List of Recommended Dosages, pp. 169–189).

If the condition still doesn't improve, visit your vet, who will ask you for a thorough history of your cat's diet and her eating and elimination habits. To rule out a possible pre-existing condition that may have caused the colitis, he will perform a stool examination and perhaps a radiograph or endoscopic test. Testing for food allergies also may be advised.

If your cat does have colitis, the first step involves treating whatever underlying factor—if any—is causing the condition. For instance, diet-induced colitis responds to a daily menu of high fiber, regularity-promoting food or a hypoallergenic diet. Parasite-induced colitis can be cured by ridding your kitty's body of internal pests. Colitis related to an immune disorder is treatable with cortisone (although long-term treatment should use other less toxic anti-inflammatory remedies).

PREVENTION: Although you can do little to prevent colitis caused by illness, you can avoid diet-induced colitis by feeding your cat a consistent, high-fiber diet. Periodic stool checks can quickly uncover colitis caused by intestinal parasites (such as coccidia or hookworms), making treatment easier.

Foul-Smelling Diarrhea, Swollen Abdomen, and More Frequent Defecation

RELATED SYMPTOMS: Bright red blood may accompany the stool. Your pet may have no appetite and be weak. She may have an oily coat and may look as if she has lost weight in the legs and face. You may see actual white, ricelike worm segments around your cat's anus or in her fecal material or longer, spaghettilike or threadlike whole worms in the stool.

POSSIBLE CAUSES: Is yours a newly adopted cat or kitten that came from a cattery, pet shop, pound, or a feral cat population? Did your pet recently spend time with another cat? Have you boarded your kitty recently? Does your cat spend time outdoors where she has access to other felines? Has your pet recently had fleas? Your cat may have picked up an internal parasite, such as **coccidia, roundworms,** or **hookworms.** These parasites thrive in a cat's intestinal tract, where they absorb nutrients from her food. Coccidia (a protozoan parasite) and roundworms and hookworms (both worm parasites) are usually caught through exposure to an infected cat's fecal material. Tapeworms are usually transmitted by fleas.

CARE: Suspect internal parasites? Gather a stool sample from your pet and drop the material off at your veterinarian's office. By analyzing it, not only will he be able to diagnose whether your pet has an internal parasite, but he can tell which parasite—or parasites—she has. Should you discover worms in your pet's stool, you may be tempted to pick up a commercial dewormer from your local pet store. Before you spend the money, however, talk to your vet. Over-the-counter products may not be adequate for the task because most commercial dewormers only rid the body of a few types of intestinal parasites, and kittens often have several different parasites at the same time. Upgrading your pet's nutritional status to a chemical-free, high-quality diet, accompanied by digestive-enzyme supplements and antioxidant vitamins C and E (see Appendix E: List of Recommended Dosages, pp. 169–189), may help your cat's immune system fend off unwanted invaders.

If a worm or combination of worms is plaguing your pet, your vet will administer a dewormer every 3 weeks until the stool sample comes back negative. In the case of coccidia, he will send you home with several weeks' worth of oral sulfonamide medication, such as Albon. To prevent taxing an already-stressed intestinal tract, your vet may recommend an easily digestible diet for a few days to a week, such as strained-meat baby food (look for brands made without stomach-upsetting onions).

PREVENTION: Do you have more than one cat? Before allowing a newly adopted cat or kitten to play with the resident feline (or felines), deliver a sample of her stool to the vet for an internal parasite check. Until the newcomer is certified as parasite-free, keep her (and her litter box) away from all other household animals. Vets recommend that anytime you acquire a cat—even if you don't already have other kitties—you should have her checked for parasites. Checking the stool of a new cat should be performed several times at 2-to 3-week intervals before concluding the cat is parasite-free.

No Bowel Movements, Thick Mucous in the Rectum, Vomiting, and Lack of Appetite

RELATED SYMPTOMS: Your cat may be lethargic and have a painful abdomen and/or a fever. She may adopt a hunched-up stance.

POSSIBLE CAUSE: Has your cat recently ingested a small object or poisonous substance? Has she been diagnosed with a hernia, an abdominal tumor, or intestinal parasites? If you answered "yes" to any of these questions, your cat may have **blocked bowels:** This condition typically occurs when a small portion of the intestines becomes obstructed (most often where the small and large intestines meet, although it could occur anywhere along the gastrointestinal tract). The ailment can be traced to a severe gut-area inflammation, a foreign object, a hernia, a tumor, or parasites.

CARE: Take your cat to the vet immediately. If left untreated, a blocked bowel can kill the intestinal tissue and, eventually, the cat herself.

To reach a diagnosis, your vet will perform both plain and dye-contrast radiographs. If there is an obstruction, emergency surgery will clear the blockage and remove any dead tissue. Post-op care includes 2 days of intravenous feeding (requiring a hospital stay) and 1 week's worth of antibiotics. To produce a good bowel movement, replace half of your cat's regular amount of food with fresh raw meat (a natural laxative) and fresh raw vegetables (to provide bulk), and mix her food with either powdered psyllium (such as Metamucil) or mineral oil 2 times a day. If these recommendations still do not produce a significant bowel movement, a fleet saline or mineral-oil enema can be given several times a day (your vet will teach you how) until one is produced. (Enemas designed for humans should *never* be given to cats—they cause dehydration.)

PREVENTION: Watch what your cat puts in her mouth.

Straining to Defecate, Regardless of Stool Consistency

RELATED SYMPTOMS: The diarrhea may or may not alternate with constipation. The cat may crouch forward and strain when defecating. The stool may have a foul smell.

POSSIBLE CAUSE: Is your cat fed "people food" or does she have access to the garbage can? Does she have internal parasites, such as coccidia or hookworms? Does she have food sensitivities? Has she been diagnosed with feline leukemia, feline AIDS, an immune disorder, a fungal infection, internal polyps, or tumors? A "yes" to any of these queries may indicate **colitis.** Colitis is an inflammation of the colon's mucous lining, which often results in painful diarrhea. The condition can be acute (a sudden, short bout) or chronic (long-term).

CARE: If the condition appeared suddenly in response to a change in diet, try withholding food from your cat for 24 hours. Slowly return to feeding your cat her original diet, or gradually, over a 2-week period, introduce a new diet of bland, hypoallergenic food—strained-meat baby food (made without stomach-upsetting onions) is ideal. Vitamin A has been found to help certain types of colitis. To soothe your pet's colon lining, try giving her aloe vera juice. (See Appendix E, pp. 169–189, for recommended dosage). You also can treat your cat's diarrhea with activated charcoal (available at your local pharmacy), which helps absorb toxins, poisons, and other irritating material.

PREVENTION: Keep your cat inside and away from garbage cans and stray cats who could be carrying disease. Feed your pet a high-quality diet and avoid giving her "people food."

Pregnancy

A female cat is sexually mature between 4 and 9 months of age. The exception to this is the Persian, who may be 1 year—or even 18 months—old before maturing sexually. This time span varies widely depending on the cat, but if she is not fixed, she will typically go through estrus (commonly known as "heat") between 4 and 7 months. If your cat is going to get pregnant, this is the time when it happens.

A vet usually can tell your cat is pregnant within 4 weeks—simply by feeling her belly. Not only is it heavier, but the internal swellings of the uterus caused by the unborn kittens can be felt. Because a feline pregnancy lasts an average of 63 to 65 days, the process is much more accelerated than a human pregnancy. Thus, visit your vet as soon as you suspect your cat is expecting. Your vet will examine your kitty and perhaps place her on a special diet.

At 45 days, she will begin eating more and more. In fact, in the last 2 weeks of her pregnancy, your kitty may be eating 30 percent more than she consumed before becoming pregnant. Although cats remain active throughout their pregnancies, stick to gentle play. Avoid any wild games of cat-and-mouse and any activity that may spark your cat to jump or leap.

During this later stage (approximately the final 2 weeks), your cat will also want to prepare a warm, secluded, draft-free nesting area where she'll give birth and later attend to her kittens. Your vet can help you by offering you guidelines and suggesting suitable nesting material, such as blankets and towels, for your particular cat's needs. Don't be surprised if, even after you've provided a superb nest, your pet seeks out a quiet spot under the bed, in a closet, or behind the couch where she will probably move her nesting materials as well.

During this time, she may shred the bedding you've provided for her, claw at the floor of the nesting area, and even tear out her own fur. She may also root around in your closets, rummage through your drawers, and paw at your bedspread. All of this is completely normal: She's trying to make the area softer and more comfortable for both herself (after all, she's got a long day ahead of her) and her soon-to-be-born offspring. Also completely normal are irritability, restlessness, and increased grooming habits.

115

About 1 week before your cat is due to deliver, she will begin to spend more time cleaning herself, particularly her abdomen and genital areas.

About 2 to 3 days before labor, your cat's breasts may enlarge. Even closer to delivery, signs that your pet will be giving birth within the next 24 hours include crying, pacing, restlessness, and disinterest in food. She may even vomit 1 to 2 times. All of these signs constitute prelabor, the period just before labor. She will spend her time waiting for labor to begin, sitting in her new nest. You may want to move her litter box within sight of her nest, although not necessarily right next to it.

At the 61-day mark (or the day the vet estimates to be the 61st day), your vet may suggest that you take your cat's rectal temperature daily. (However, some vets feel this unduly stresses an already-stressed cat. Let your pet and your vet be your guides.) Between 12 and 24 hours before she is due to deliver, the rectal temperature may drop from its normal mark around 101.5°F to 99.5°F or lower. (However, this 2-degree drop isn't universal, and some cats' temperatures remain normal.)

As the birthing day approaches (your vet will help you estimate when this should be), keep your cat indoors. The feline world is ripe with stories of pregnant kitties running off to deliver their babies in a nearby barn, a neighbor's garage, or an abandoned car. Needless to say, your home makes a safer maternity ward than any of these.

The birthing process is a wondrous experience, so it's tempting to ask people to share it with you and your pet. However, remember that this is a physically and mentally challenging time for your feline friend. She's in pain, she's working hard, and she doesn't need the added emotional strain of having strangers—or, for that matter, too many human family members—watching the action. Someone should certainly watch her progress, but limit the number of witnesses to one predesignated human. No one but this individual should be allowed near the birthing mother. If your cat indicates in any way that she wants to be alone, this person should leave the room—though you (or whoever the designated human is) can occasionally peek in on your pet to make sure everything is going well.

This point can't be repeated enough: Your pet's emotional and physical comfort should be your utmost concern. It's not unusual for cats to delay the birthing process—or even interrupt already-started birthing—for several days if they are uncomfortable with any activity or person (even you, her beloved owner) in the household. This behavior is rooted in the wild, where a birthing mother must be ready to stop what's she's doing and find safer ground should a threat suddenly appear.

If prelabor has continued for 36 hours and you have yet to see a kitten—or even a contraction—*call your vet*. Your cat may need help in birthing her kittens, or may be uncomfortable in her surroundings. Either way, professional help may be necessary.

You'll know labor has begun when the cat begins to strain and you see contractions. You may also notice a slightly greenish-tinged or straw-colored secretion excreted from the vulva. This fluid is passed after the amniotic fluid around the first kitten is ruptured, and a kitten usually follows within 30 minutes. (However, don't be alarmed if the rupture does not occur—it isn't universal to feline births.)

Contractions are strong, visible, straining movements that involve both the abdominal muscles and diaphragm. During active labor your cat may either lie on her side or her sternum (chest), or she may squat (a position resembling defecation). In normal situations, the first kitten will be born within 2 hours of the onset of labor. Again, if you see no kittens, *call your vet*.

In a normal birth, babies emerge 15 to 30 minutes apart. Most newborns appear with their front feet and noses first. Your cat will eat the membrane that encases the newly born kitten, then chew off the umbilical cords. She'll then give each baby a good lickdown: Not only does this clean the kittens, it stimulates their circulation and encourages them to nurse.

If, for some reason, your cat does not eat the encasing membrane, you will have to break open the sac so the kitten can breathe. Use your fingers and gently tear the membrane away. Should the mother fail to sever the umbilical cord, you can wait until after she's done delivering all her kittens to see if she will get around to it. If she doesn't, get a *clean* piece of sewing thread or unwaxed dental floss and tie it around the cord an inch from the kitten's body, then cut (using a disinfected pair of scissors) or break (using your fingers) the cord just past the thread or floss. A *word of warning:* If the cord is severed too near the kitten's navel, or cut too cleanly, it might continue bleeding. Should this scenario occur, clamp or pinch off the cord and tie a second thread around the stump. No matter how it's removed, the umbilical stump should be disinfected with an antiseptic, such as iodine. However, make sure the mother cat does not lick at the stump while the iodine is still moist (2 to 3 minutes after iodine application should suffice).

A **placenta**, or **afterbirth** as it is also called, should follow each kitten. This is not the same as the **protective membrane**, called the **amniotic sac**, which encompassed the newborns. Your cat may eat each placenta, she may eat a few, or she may eat none. (In fact, it's probably better if she doesn't consume all the placentas; she could end up with a bad case of diarrhea.) What should concern you is that a placenta emerges after each kitten. If there are five kittens and you've only counted four placentas, one of the placentas is trapped in the birthing canal and needs to be removed. If left, it can cause infection, or hamper the birth of any remaining kittens.

If you notice that a just-born kitten isn't breathing, hold it upside down and massage its skin. This allows any trapped fluid to drain from its lungs. If this isn't working, use a bulb syringe and suction out excess fluid that may be stuck in the mouth and nose. If, after 2 minutes, the infant still isn't breathing, you will have to try artificial respiration.

Start by opening his mouth and lying the tongue to one side between the top and bottom molars. Place your finger in the kitten's mouth and throat to feel for any obstructions, including vomit, mucous, or blood. Remove anything you find. Extend the kitten's head and neck, then close his mouth. Inhale deeply, completely cover the kitten's nose and mouth with your mouth, then exhale into his nostrils. When working with such a tiny creature, be gentle—you don't want to damage the nostrils or overinflate the lungs. The air should reach his chest: Watch for the chest to swell. Remove your mouth and allow the kitten's chest to deflate normally. When it has, put your mouth over his nose and mouth and start again. This inflate-deflate cycle should be done 12 times per minute, until the kitten begins breathing on his own.

Often, immediately after a kitten has ceased breathing (or just prior to it), his heart may stop. (You can check for his heartbeat by wrapping your hand around his chest, just behind his front legs, and applying slight pressure. Alternately, you can check the kitten's pulse, which is best felt on the inner side of the thigh in the groin area.) If the nearest vet is some distance away, you also will have to perform external cardiac compression. Place the kitten on his right side, laying him on the firmest surface possible. Place your thumb on the side of his sternum (chest) that is facing up and wrap your other four fingers around the chest so that your fingers are underneath his body and pressing on the other side of the sternum. Squeeze the chest firmly between the thumb and four fingers. Release. Then squeeze firmly again. Release. Repeat 6 times, then wait 5 seconds to see if the chest expands and whether any pulse or heartbeat results. If it doesn't, repeat. When you combine pulmonary resuscitation with external heart massage, you should perform 1 pulmonary expansion for every 6 cardiac compressions.

After giving birth, your cat will rest and nurse her brood. While her appetite may take a few long hours to return, she may want water.

Dark Red, Brown, Yellow, or Black Fluid Excreted from the Vaginal Canal with No Following Birth and/or More than 30 Minutes Between the Emergence of Each Kitten; or 60 Minutes of Intense Straining without the Birth of a Kitten; or 10 Minutes of Intense Labor with a Kitten Visible in the Birth Canal

RELATED SYMPTOMS: This liquid excreted from the vaginal canal may contain pus and/or smell foul. Your cat may strain needlessly and show an excessive amount of weakness or discomfort. Although this last symptom can be difficult to gauge (since giving birth is an extremely uncomfortable situation) be sensitive to crying, growling, or vomiting. You may notice the kitten's head or other body parts appear at the vaginal opening during a contraction, then slip back into the birth canal once the contraction has ended.

POSSIBLE CAUSE: Is your cat under a year old? Is she elderly? Is she overweight? Is she in generally poor health? Has she, at some point in her past, suffered a pelvic injury? Is she a Persian or Himalayan? Your cat may be suffering from **dystocia:** In other words, she's having a difficult labor. Although not as common in felines as in canines—or humans, difficult births do happen in the cat world. The condition—which usually appears with the first kitten—can be caused by a kitten or afterbirth stuck in the birth canal; kittens whose heads are too big for your cat's birth canal (especially common among Persians and Himalayanas due to their broad, domed foreheads); a kitten positioned awkwardly; or uterine inertia. Basically the uterus becomes exhausted and too tired to continue contracting. Dystocias are most often due to especially large infants and/or a particularly small birth canal.

CARE: First be sure that your pet hasn't interrupted her own labor in response to something you or a family member has done, or too much activity elsewhere in the house. If this is the case, your cat probably isn't experiencing dystocia and won't have any of the physical symptoms just listed. She is simply waiting until "the coast is clear" before bringing her offspring into the world. When feeling uncomfortable with her environment, a feline can put heavy labor on hold for up to 24 hours.

If you can't blame the environment and physical signs indicate a problem, **call your vet**. He may make a housecall, ask you to rush your cat over for an emergency cesarean, or talk you through the process over the phone. Typically, here's what you might be told: Wash your hands with antibacterial soap or dish-washing soap, then lubricate a finger with petroleum jelly and insert that finger into the vaginal canal. You should feel a kitten. Can you figure out where the head and front and rear legs are? Often a kitten's leg will bend under itself, causing it to become stuck. If that feels like it may be the case, you can reposition the leg. Remember, for your kitty's comfort, *be gentle!*

If the baby is well into the birth canal and seems to be in a normal position, you can help ease it through the canal. Using your thumb and forefinger, gently grasp the animal around the shoulders (if you can reach them). You don't want to put pressure on her head, neither should you tug on the amniotic sac surrounding the kitten. Wait for your cat's natural contractions. When one strikes, softly pull the kitten downward (the direction the vagina is angled). If the kitten's head seems too large to fit through your mother cat's vulva, there's a chance you can very gently finesse the edges of the vulva to fit around the kitten's head.

If, after all your efforts, your cat is unable to deliver her kitten, you will need professional assistance. Your vet may be able to physically manipulate the kitten inside the birth canal to make it easier for the baby to pass through, or he may give the cat injections to make the contractions stronger. Be aware, however, that stronger contractions mean that your already-suffering cat will be put through more pain, which explains why many vets avoid this latter option. If neither option works, your vet may need to perform an emergency episiotomy (an incision to enlarge the opening of the vagina) or a cesarean.

In some cases, a retained placenta is blocking the delivery of the next kitten. Using the same steps just outlined, you can often find this afterbirth. Grasp it, and gently but firmly pull until it passes out of the vaginal canal. After this, your cat may not be ready to deliver the next infant right away, or she may need a prolonged rest period before the next kitten is delivered.

PREVENTION: Cats who have one difficult birth tend to repeat the experience the next time they get pregnant. Prevent this painful scenario by having your cat receive an ovariohysterectomy. Or, if your cat is less than a year old, elderly, overweight, in poor health, a Persian or Himalayan breed, or has suffered a pelvic injury in the past, think twice before letting her become pregnant in the first place. Keep her inside if she has not been spayed.

Sex Organ Disorders

Unfixed cats of both sexes are more likely to suffer from infectious diseases and other physical conditions involving reproductive organs. Why? Because these unneutered animals have more organs to be affected. The reproductive parts of both sexes can be strained, injured, or infected by a number of conditions related to sexual activity. In addition, females can encounter many complications during heat, pregnancy, and lactation.

Only a portion of your cat's reproductive system is external. Called genitalia, these sexual parts are easily monitored for physical changes. Perhaps a testicle will look swollen, or the vulva will be inflamed. (Although breasts are not categorized as genitalia, a breast that is swollen or hard to the touch is not normal and may be a useful symptom to consider when making a diagnosis for mastitis or a tumor.)

Then there are instances when the genitals and/or breasts will look normal, or perhaps be only slightly swollen, but will excrete some type of discharge. This material can be milky, clear, greenish, or yellowish. It may or may not contain mucous, blood, or pus—in very large or quite small quantities. Other than the cat constantly licking the discharge away, the only sign that something is amiss with the interior reproductive organs may be this discharge.

In females, a vaginal discharge often signals **pregnancy-related problems,** such as **placental disease, the uterus's failure to repair itself after delivery,** a **uterine infection**, or simply that the cat is **in heat.** In males, **trauma to the genitals, growths in the penis,** and **prostate disorders** are the leading causes of discharge. To determine exactly what is causing the discharge, however, your vet will physically examine your pet, study the discharge's content microscopically, and take X rays if indicated.

Note that veterinary professionals routinely urge pet owners to get their female and male cats neutered. Not only are neutered pets less prone to ailments involving the reproductive organs, they are less likely to exhibit sexually based behavior problems, such as excessive roaming or aggression. Another very good argument for having your female or male cat fixed is the staggering number of unwanted cats in America.

Each day, hundreds of thousands of cats are either abandoned by owners who no longer want them or are put to sleep in overcrowded shelters. In order to avoid contributing to this very preventable tragedy, keep your pet from breeding. If you really want your cat to have a kitten in her life, make sure you can find good homes for all of the kittens or consider adopting one of the many already-abandoned newborn felines found in a local shelter.

Blood-Tinged Discharge from The Vulva Several Weeks After Giving Birth

RELATED SYMPTOMS: Thin, blood-tinged discharge is normal during the first 3 weeks after giving birth. The passing of bright blood or any excretions occurring more than 1½ months after delivering is not. The passage of pus at any time is not normal, and the cat should be seen by a vet immediately.

POSSIBLE CAUSE: Has your cat given birth in the last 3 weeks? It's possible that she has a condition called **uterine subinvolution**. In layperson's language, this means the uterus is not properly repairing itself after the birthing process. The condition can lead to anemia and uterine infections.

CARE: Take your cat to the vet, who will make a diagnosis after evaluating your cat's physical condition. If the bleeding is mild, the uterus may simply be taking its time repairing itself and will finish healing in another 2 to 3 weeks. In extreme cases, such as severe infection, the uterus may have to be removed. Persistent bleeding may warrant iron supplements to treat or ward off anemia; your vet can prescribe these. If an infection is present, antibiotics will also be prescribed.

PREVENTION: Spay your cat.

Excessive Licking of the Genitals, Vaginal Discharge, and Dragging the Rear End Along the Ground

RELATED SYMPTOMS: The discharge may contain blood, pus, or mucous. The outer folds of the vagina will appear large and swollen, and the inner vagina may protrude. You may also notice appetite loss, lethargy, vomiting, increased water intake, fever, abdominal pain, and weakness in your cat's hind legs.

POSSIBLE CAUSE: Is your pet sexually active? Has she recently given birth? Has she been diagnosed with a metabolic disease such as diabetes, or is she being treated with estrogen-based medication? Any of these can point to an **infection of the reproductive tract,** usually the vagina or uterus. Of course, keep in mind her normal reproductive cycle and consider the possibility that she may be in heat.

CARE: Take your pet to the vet, who will microscopically examine a stained slide of the cat's discharge and possibly take a bacterial culture and/or X rays to reach a diagnosis. If infection is present, antibiotics will be prescribed, which can either be applied via a douche infusion (by your vet), administered orally, or injected. If dehydration is present—this is more common with an infected uterus—intravenous fluids also will be given. If there are any nursing kittens, they will be transferred to a formula diet.

A severely infected uterus in an unspayed female will prompt your vet to recommend spaying. This is essentially an ovariohysterectomy, which involves removing the ovaries and the uterus. The rationale? Spaying your pet will prevent the condition from recurring—an important reason, since uterus infections often recur and can lead to liver and kidney damage, or even a ruptured uterus and peritonitis.

PREVENTION: Since many reproductive-tract infections are caused by sexual activity, consider having your cat neutered or strictly confining her during her heat periods.

One or More Lumps or Flat Patches on a Breast

RELATED SYMPTOMS: These might be ulcerated. If one lump is noticed, check all other mammary glands on both sides for more lumps.

POSSIBLE CAUSE: Is your cat a Siamese? Is she a female older than seven? Has she had two or more pregnancies (or false pregnancies), or has she ever received hormone shots to prevent estrus? Is she unspayed? If you can answer "yes" to any of these, your kitty may have **breast tumors.** Breast tumors are the third most common types of tumors found in felines. Unfortunately, the majority of them are malignant.

CARE: The earlier that breast tumors are found and removed, the better the chance for survival, making immediate vet care the most prudent path.

To reach a diagnosis, your vet may take a tissue biopsy or remove the entire lump to learn whether the lump is malignant. This is important because breast tumors can be confused with a condition called mammary hyperplasia. The latter ailment occurs when the mammary gland develops too many cells, making the breast look larger than normal. Mammary hyperplasia is usually noticed in young, unspayed cats around the time of their first estrus, usually between four and seven months of age. The condition can be treated by spaying the cat.

If the condition is a tumor, and if that tumor is malignant, an X ray will determine whether the cancer has spread to the lungs. If it has, it is too late to operate; your vet will discuss how long your cat may live and how to make her remaining time comfortable. If the cancer is limited to the breasts or the lumps are benign, the growths and surrounding tissue will be removed. (Because they can become malignant, even benign lumps should be removed.) After surgery, be sure that your cat does not scratch, paw, or bite her incision: An Elizabethan collar may help to restrain your kitty from touching the incision until after it has healed and the sutures are removed.

PREVENTION: Spay your cat as early as possible. Research has shown that cats spayed before their first heat have the lowest incidence of breast tumors. Unspayed female cats are seven times more likely to develop mammary tumors than spayed females.

Painful, Swollen, Warm Breasts in a Cat with Kittens

RELATED SYMPTOMS: The condition may exist in one, some, or all breasts. A watery, puslike, or even bloody secretion leaks from the nipples when pressure is placed on the breasts. The breasts may feel firm or even hard. The cat frequently has a fever.

POSSIBLE CAUSE: Are your cat's kittens rough when nursing? Does she have trouble releasing milk from all nipples? She may have **mastitis**, an inflammation and/or infection of the mammary glands that is somewhat uncommon in cats. The condition is caused when glands are damaged by coagulated milk or injuries inflicted by hungry kittens.

CARE: If there is no bloody discharge and the area is only slightly swollen, massage the affected breasts with rubbing alcohol or a mentholated camphor ointment (available at a drugstore) to increase circulation and lower the skin temperature. You can also apply hot packs and express the gland to draw out some of the coagulated and caked milk. Do not let the kittens nurse on the affected glands. (You can prevent them from doing so by placing adhesive tape or a bandage over the infected nipples.)

If blood is being secreted, or if the breasts don't improve after 48 hours of homecare, take your cat to the vet, who will examine her. If the breasts are abscessed, your kitty will be put under anesthesia so that the vet can lance, drain, clean, and bandage the glands. Antibiotics will be given and the kittens placed on a formula diet.

PREVENTION: Check your nursing cat's breasts daily. Because an infection can be passed on to kittens, any change in breast conditions should be addressed immediately and the infants placed on formula.

Redness and Inflammation of One or Both Testicles and a Stiff Gait with Straddled Hind Legs

RELATED SYMPTOMS: The scrotal area will be painful to the touch, and the scrotum may be abscessed and filled with foul-smelling pus. The cat may have a fever and/or refuse to eat.

POSSIBLE CAUSE: Does your cat spend unsupervised time outside? Has he recently suffered a groin injury, scrape, or bite to the testicles or scrotum? A "yes" to either of these questions could indicate **orchitis**, also known by the less medical-sounding **inflammation of the testicles**.

CARE: If the area is simply swollen and red without the presence of abscesses, you can attempt home treatment. Prepare a large batch of chamomile tea and let it cool to room temperature. Gently swabbing the area 3 times daily with this liquid can help keep the inflammation in check.

If there are draining abscesses, or if you see no improvement in 48 hours of homecare, take your cat to the vet, who will prescribe antibiotics and lance the scrotum if it is abscessed. If your kitty can't stop licking himself, your vet will fit him with an Elizabethan collar. This plastic, funnel-like contraption is worn around the neck and, if properly sized, makes it virtually impossible for your pet to reach his backside. If the condition is severe enough, castration may be recommended.

PREVENTION: Supervise your cat's time outdoors. To ward off infection, treat any injuries, even small cuts and scrapes, immediately. If you are not interested in breeding your cat, consider castration.

Sexual Behavior Abnormalities

Not all sexual conditions have physical symptoms. You may notice your cat acting differently, thus signaling something is wrong with part of the reproductive system you can't see. A well-known example is the oversexed male who—much to his owner's horror—aggressively attacks other animals, people, or even inanimate objects. No, this isn't everyday male cat behavior. It's caused by too many male sex hormones, and it can be treated, as can other abnormal sexual behaviors, including roaming (males) and mothering toys (females).

As with symptoms involving the genitals, sexual-based behavioral symptoms are much less common in cats who have been neutered. The youngest age a cat should be neutered was once thought to be 6 months. However, many humane societies and veterinarians are now neutering as young as 8 to 12 weeks. In addition, research has shown that neutering cats before they reach 1 year affords the greatest protection from mammary cancer, reproductive-tract cancers, and uterine diseases in females.

Aggression, Swiping, or Biting While Being Petted

RELATED SYMPTOMS: Your cat may wander excessively and constantly mark territory.

POSSIBLE CAUSE: Is your cat an unneutered male? There's a good chance he is displaying an aggressive tendency as a result of an **overproduction of male hormones.**

CARE: Neutering is the ultimate solution. If, for some reason, that isn't an acceptable option, your vet can try to temporarily keep your cat's aggression in check by prescribing a female hormone.

PREVENTION: Neuter your cat, or at least keep him confined so that he cannot breed with neighborhood females in heat or have territorial fights with the males.

Defending of Toys by a Nonpregnant, Unmated Female, Nest-Building, and Enlarged Breasts

RELATED SYMPTOMS: The abdomen may also appear enlarged. If pressure is applied to the breasts, the nipples secrete milk. You may also notice aggressive behavior, a tendency to hide in quiet places, and a change in appetite. Most symptoms are obvious 40 to 45 days after your pet has ovulated.

POSSIBLE CAUSE: Was your cat in heat a 6 weeks ago? Are you positive that she has not mated during an unsupervised moment? Although the condition is rare in cats, she may be experiencing false pregnancy. After ovulation, a cat's progesterone levels remain high for about 45 days, even if she has not mated. These higher hormone levels lead some females to feel—incorrectly—they are pregnant. To further confuse things, their body functions and behavior seem to scream "pregnant."

CARE: There is no way to tell false pregnancy apart from the early stages of real pregnancy simply through physical observation. Unless you are absolutely, positively, 100 percent sure that your cat has not slipped out and mated, there's the chance that she is pregnant. If you know she's not, you can only wait patiently until she "snaps out of it." This should take no longer than 2 weeks.

Milk production is rare in feline false pregnancy, but if your cat's breasts do become swollen, massage them with diluted alcohol to stimulate circulation to the area and reduce swelling. Keep her distracted by extending her playtime. Remove all nesting materials, such as blankets, pillows, and towels. If she's adopted any toys or socks as surrogate kittens, take those away, too.

If her breasts remain inflamed or she's still behaving this way after 2 weeks (or you're unsure whether she's expecting), take her to the vet. An X ray (after 45 days of pregnancy) or diagnostic ultrasound can reveal the presence—or nonpresence—of kittens. Your vet may recommend a hormone injection to help shorten the false pregnancy.

PREVENTION: Spay your cat.

Nervousness, Pacing, Whining, and Trembling in a Pregnant Cat or New Mother

RELATED SYMPTOMS: Your cat may have a fever and respiratory problems. She might appear uncoordinated, have muscle spasms, and pant. The condition could be mistaken for an epileptic seizure.

POSSIBLE CAUSE: Is your cat due to have kittens in the next 2 weeks? Or has she given birth within the last 3 weeks? Is she undernourished? She may be suffering from **eclampsia**, a rare but serious condition caused by reduced blood-calcium levels. The illness typically strikes poorly nourished cats and cats who have recently given birth to large litters.

CARE: Take your pet immediately to the vet: the accompanying exhaustion, high fevers, and respiratory troubles can lead to death. Your vet can reach a diagnosis by observing the foregoing clinical signs and analyzing the cat's blood for low calcium levels. Fortunately, eclampsia is easily treated with intravenous calcium and cortisone injections. If your cat has kittens, they will be placed on formula, so that they no longer steal the mother's calcium.

PREVENTION: If your pet is pregnant or has just given birth to an unusually large litter, talk to your vet about calcium supplements—this is especially important if she is a small cat.

Vomiting

In humans, vomiting can be a frightening experience. Not only is it extremely uncomfortable, it usually accompanies some kind of problem, such as a flu, bulimia, a high blood-alcohol level, motion sickness, food poisoning, or an overdose of medication. In cats, however, vomiting isn't always a cause for alarm. Kitties often throw up from ingesting hairballs, snacking on grass, accidentally ingesting a toxic substance, or chowing down more dinner in a shorter period of time than their stomachs can handle.

How do you distinguish "neutral vomiting" from vomiting that signals a medical condition? Severity. Vomiting small amounts of hair or bile 1 to 2 times a week may not be unusual for a longhaired cat. Vomiting continually for 1 hour straight, on and off for 24 hours or more, or 1 or 2 times every day is of greater concern. Besides indicating a medical condition, the vomiting itself can cause an additional problem of its own: dehydration.

The presence of other symptoms helps your vet diagnose your pet's condition. When containing blood or mucous, vomiting usually indicates **gastrointestinal disease** or possibly the presence of a foreign body in the intestines. When joined by severe abdominal pain, **pancreatitis** or **urinary-tract disease** may be the cause. When the vomiting is accompanied by frequent visits to the litter box with straining but no urine production, **feline urologic syndrome** should be suspected (see the section, Straining to Urinate, but Producing No Urine, Crying, Attempting to Urinate in Places Other Than Litter Box, and Licking Genitals (in Male Cats), pp. 107–108).

This is a good time to mention what vomiting is not: regurgitation. Vomiting is the forceful ejection of food and/or fluid from the stomach or the upper small intestine. Regurgitation is passive; food "just comes up"—not from the stomach or duodenum as with vomiting, but from the esophagus. Because they have different sets of causes, distinguishing between the two is important.

Chronic Vomiting and/or Diarrhea That Varies in Intensity with Change of Diet

RELATED SYMPTOMS: Your pet has intestinal gas and intestinal sounds and possibly mucous in the stool. Her bowel movements are more frequent than usual. In addition to gastrointestinal symptoms, skin-allergy symptoms, such as inflammation, itching, and scratching, may appear.

POSSIBLE CAUSE: Did your kitty begin displaying these symptoms at less than six months or over six years of age? Does the problem seem to remain at the same level of intensity regardless of the season? Your cat may have a **food allergy** or **hypersensitivity**. Food allergies are thought to develop as the result of ingesting a food substance that causes an immune-system reaction in the pet's intestinal tract or bloodstream. (Food intolerances or sensitivities are not true allergies, but are close enough to be treated as such.) Certain types of food hypersensitivities are sometimes referred to as nonimmune food intolerance and are thought to be caused by chemicals with high levels of inflammation-producing histamines.

CARE: Food allergies can be diagnosed in two ways: blood testing using the RAST or ELISA test and restricting the diet. An elimination diet works by restricting your pet's food to a single protein and a single carbohydrate food source, and then gradually introducing additional foods until the culprit is discovered. Of course, once the offending substance has been identified, you will want to eliminate it from all of your pet's meals.

PREVENTION: Feed your pet a chemically free diet. Avoid all food ingredients that appear to be allergenic in your pet's blood test or during the elimination/provocation process. Antioxidant therapy with the vitamins C, E, A and the mineral sulfur can help minimize symptoms (see Appendix E: List of Recommended Dosages, pp. 169–189).

Repeatedly Vomiting Frothy, Yellowish Bile (Especially After Drinking Water), Accompanied by Appetite Loss, Depression, Fever, and Diarrhea

RELATED SYMPTOMS: Although your cat may seem to want more water than usual, she is unable to drink it. You may find her crouched in pain over her water bowl, her head hanging a few inches from the surface. Her abdomen will be painful, causing her to cry repeatedly. The diarrhea may be yellowish and/or blood-streaked.

POSSIBLE CAUSE: Does your cat spend time outdoors where she can come into contact with racoons and/or other felines? Is she a stray or did she come from a shelter or humane society? Has she been exposed to the litter boxes and/or food bowls of·other cats? She may have **feline panleukopenia,** also known as **feline infectious enteritis, feline parvovirus,** and **feline distemper** (although the feline panleukopenia virus is in no way related to the virus that causes distemper in dogs).

Feline panleukopenia is one of the most serious and widespread viral diseases in cats, and is a leading cause of infectious-disease deaths in kittens. The virus encourages disease by destroying your cat's white blood cells, thereby hampering her body's ability to fight off the disease.

The virus is highly contagious and spread numerous ways: by direct contact with an infected cat or raccoon or its feces; by inhaling any infected secretions that become airborne when a sick animal sneezes; by exposure to contaminated surfaces and utensils, such as food pans and litter boxes; and via the clothes or hands of people who have touched an infected animal, such as personnel at a breeding facility. The bite of an infected flea or other external parasite also can transmit the disease. Unfortunately, the distemper or FPL virus (as feline panleukopenia is also known) is extremely hardy. Not only can it live in carpets, cracks, and furnishings for 1 year or more, it's resistant to ordinary household disinfectants.

Once your kitty has been exposed to the virus, the symptoms follow rapidly. An animal usually shows clinical signs of the illness within 2 to 10 days of exposure. Often the most severe symptoms are not produced directly by the panleukopenia virus itself, but by a secondary bacterial infection that results from the viral destruction of the body's white cells and immune system.

CARE: FPL can be fatal: **Take your cat to the vet immediately.** If the condition is addressed early, your cat has a chance of surviving. Because feline panleukopenia can be mistaken for poisoning, your vet will take a careful history and perform a white blood cell count to determine whether FPL is present. Treatment is supportive, and includes fluid replacement to prevent dehydration, antivomiting drugs, and antibiotics to keep any secondary bacterial infections at bay. Because the virus is so contagious, be sure to keep your infected cat away from other cats. (Even if they have been vaccinated, I recommend playing it safe.)

PREVENTION: A vaccination is available and should be given to all kittens. Adult cats need yearly boosters. Keeping your healthy cat indoors will lessen the chances that she can come in contact with infected animals.

Vomiting, Dazed Demeanor, and Increased Thirst

RELATED SYMPTOMS: The vomit may be bloody and might occur more frequently after drinking. It may come on suddenly and violently or be more sporadic. Your cat may feel pain in her abdomen, and you may hear a distinct rumbling from her stomach.

POSSIBLE CAUSE: Does your pet have a food allergy? Has she eaten spoiled food or ingested a poisonous or irritating substance or foreign body? Does she ever get into the garbage, where she can find plastic, aluminum foil, paper, or other foreign material that she might have swallowed? Does she ever catch lizards and bite off their tails? Does she eat grass often? A "yes" to any of these questions might signal **gastritis,** an inflammation of the stomach lining.

Gastritis can occur either suddenly—say, after ingesting an irritating substance—or chronically. When the vomiting comes on suddenly, the condition is called **acute gastritis,** and usually occurs when the cat swallows an irritant (such as grass) or a poison that the stomach tries to expel immediately.

Chronic gastritis is typically due to hairballs, plastic, aluminum foil, or another foreign object that sits in the stomach for some time. Cats with chronic gastritis may vomit only small amounts at a time, usually sporadically. The vomit sometimes contains food from the previous day. Cats with the chronic condition also may appear lethargic, have a dull coat, and lose weight. Vomiting may also result from infectious disease of the GI tract or other parts of the body. Kidney, liver, pancreatic, and uterine disease may also trigger vomiting.

CARE: If vomiting is constant, **take your cat to the vet immediately**. If the vomiting isn't constant, try withholding food for a full 24 hours. Your goal is to rest the stomach.

After passing the 24-hour mark, allow your pet to have water. If no vomiting follows, try a 2-day diet of easily digestible, strained-meat baby food (look for a brand made without stomach-upsetting onions). Divvy this into 4 small daily meals (3 to 6 ounces each, depending on your cat's size and appetite). Provide small amounts of water (preferably Gatorade or sugar water, which is made by combining 3 parts water with 1 part maple syrup) frequently, but prevent consumption of large amounts of fluid all at once, which may stimulate vomiting. If the water and small meals are tolerated, your cat can return to her regular diet.

If, during this time, there is no improvement, take your cat to the vet, who will perform a physical exam and laboratory tests, and take X rays. With these tests, he will try to rule out conditions such as liver disease, kidney failure, hyperthyroidism, heartworms, diabetes, tonsilitis, pancreatitis, or an infected uterus.

Should **primary gastritis** (gastritis not resulting from disease in other organs or parts of the body) be the culprit, your vet will begin by determining its cause. If your cat ingested poison, your vet will follow the necessary procedures, depending on the type of toxin if that can be discerned. If your cat swallowed a hairball, your vet will administer a hairball laxative, such as Laxatone or Petromalt, to help it pass through the body. Your vet also may suggest intravenous administration of a electrolyte solution (requiring a hospital stay) to prevent dehydration caused by repeated vomiting, and recommend giving antispasmodics and an antacid to neutralize or stop the production of gastric acid.

PREVENTION: Keep a close watch on your cat's diet. Consider serving 3 or 4 smaller meals instead of 1 or 2 larger ones. A hypoallergenic diet may be in order—your vet can recommend a suitable brand of food.

Vomiting, Diarrhea, Excess Salivation, and Pain on the Right Side of the Abdomen Directly Behind the Rib Cage

RELATED SYMPTOMS: The cat shows no interest in food and appears depressed. To alleviate pain, your pet may strike a prayer pose: bent front legs and an elevated rump.

POSSIBLE CAUSE: Is your kitty middle-aged and overweight? Does she eat a high-calorie diet and/or an abundance of table scraps? Is she diabetic? A "yes" to any one of these questions might indicate **pancreatitis**, an extremely painful ailment marked by overproduction of certain digestive enzymes, which begin to damage the pancreatic tissue. (Note: While diabetes may be a symptom of pancreatitis, it may also be a result of it in extreme cases.)

CARE: In extreme cases, pancreatitis can be so excruciating that pain-induced shock and death result. In other words, **take your cat straight to the vet.** To determine whether she has pancreatitis and not an intestinal obstruction—both of which feature similar signs—your vet will perform an abdominal radiography and run blood tests to measure the level of pancreatic enzymes.

If pancreatitis is the problem, your vet will order all food and water to be discontinued for up to 72 hours. This fast will lower the number of digestive enzymes that the pancreas manufactures. To prevent dehydration, your cat may be given intravenous fluids (which requires a hospital stay), while special medication will reduce pancreatic secretions that can cause pain and destruction of the pancreas itself.

Unfortunately, once a cat comes down with pancreatitis, there's a good chance of periodic recurrence. As a preventative measure, your vet will place your cat on an easily digestible, extremely low-fat daily diet that won't tax the pancreas.

PREVENTION: Feed your cat a low-fat, high-fiber, chemical-free diet supplemented with antioxidants such as vitamins E and C (calcium ascorbate) (see Appendix E: List of Recommended Dosages, pp. 169–189). Encourage her to get plenty of exercise. Never give her fatty "people food," and don't let her become overweight. Never give your pet chicken or turkey skin, and keep the garbage can securely covered or out of sight (and smell).

Vomiting, Diarrhea with a Light-Colored Stool, Fatigue, and Dark Yellow to Brownish Urine

RELATED SYMPTOMS: The cat may have a fever and her skin, gums, and whites of eyes may be yellow-tinged. She will lose weight and have no interest in food, but may be excessively thirsty. Her abdomen may appear bloated and she may stagger and have seizures. She may experience diarrhea and/or vomiting and suffer from spontaneous, pinhead-sized hemorrhages on the gums.

POSSIBLE CAUSES: Does your cat pal around with neighborhood kitties? Could she have ingested poisonous materials, including garden pesticides or household chemicals, or picked up a blood parasite when not under your supervision? Is she taking one or more types of oral medication? Does she have heart disease, diabetes, or cancer? Has she ever experienced starvation? Has she been diagnosed with an infectious disease such as feline leukemia, feline infectious peritonitis, or toxoplasmosis? Is she extremely overweight? A "yes" to any of these dissimilar questions may indicate a **liver disease**, including **infectious hepatitis**

(also known as **acute hepatitis), chronic hepatitis,** a **tumor in the liver, hepatic lipidosis,** or **liver destruction** resulting from drugs, toxins, or parasites.

The liver filters all toxins from the blood and metabolizes carbohydrates, proteins, and fats. When asked to work overtime—for instance, when an overload of toxic substances, medications, or fat must be removed from the blood—the liver can become damaged and enlarged.

As in the case of acute or infectious hepatitis, the cat can actually catch a liver ailment from a feline pal. Another common cause of liver disease in cats is **hepatic lipidosis** (also known as **idiopathic hepatic lipidosis or IHL**). Although it strikes overweight cats most often, the exact cause of hepatic lipidosis is not known. A cat with the condition loses its appetite and shuns food. A few days, or even weeks, may pass before the animal begins eating again. Why would this "feline fasting" cause liver disease? During starvation, fat actually accumulates in the cat's liver cells, turning the liver yellow, greasy, and enlarged.

CARE: Take your cat to the vet, who will use any of the following tools to reach a diagnosis: a radiographic exam to highlight an enlarged liver, blood tests to uncover infection and/or an unnaturally high number of liver enzymes, a liver function test, or even a biopsy.

If the disease is caught early enough and successfully treated, your cat's liver may be able to regenerate itself. Thus, your vet will concentrate on eliminating the underlying cause so the organ can heal. He will prescribe a special, limited diet for your cat that contains reduced amounts of fat and protein. Depending on the condition and what caused it, your cat may be placed on antibiotics, anabolic steroids, and/or lipotropics to help move fat out of the liver cells. Nutritional support in the form of vitamins, enzymes, and antioxidants also is advisable. If a tumor is diagnosed, surgery may be necessary to remove it.

PREVENTION: Feed your cat a low-protein/low-fat diet and closely supervise her play. Do not allow her any type of access to household and garden chemicals. Be sure to use nontoxic flea products. You may want to reevaluate any long-term drug therapy your cat is taking: Certain drugs may be toxic to the liver when taken for prolonged periods of time. To avoid undue stress to the liver, promptly attend to any illness. Vaccinate your cat against infectious hepatitis and feline leukemia. Because prolonged fasting can dispose a cat to liver disease, call your vet if your pet refuses food for more than 2 days.

Vomiting, High Fever, Rapid Breathing, Arched Back, and Cautious Movements

RELATED SYMPTOMS: The cat is weak, her pulse is racing, and she may have severe abdominal pain that makes her ease into a lying-down position.

POSSIBLE CAUSE: Could your cat have swallowed a foreign object? Has she recently been injured? Was she recently diagnosed with a uterine infection or acute inflammation of the pancreas? A "yes" to one of these questions can indicate **peritonitis**, a severe condition marked by an infection of the abdominal cavity and its lining (called the peritoneum).

The peritoneum is a smooth, transparent membrane that lines the abdomen, the cavity many organs call home, including the intestines, bladder, stomach, and pancreas. If one of these organs becomes injured or torn—due to infection, a lodged foreign object, or an injurious blow—this lining can react by becoming inflamed and infected, thus affecting the entire abdominal cavity.

CARE: Peritonitis is very serious and can lead to death. **Immediately take your cat to the vet,** who will physically examine the animal as well as perform blood tests, a urinalysis, and X rays. Should your kitty have peritonitis, your vet will treat her with high doses of antibiotics and antiinflammatory medication before actually opening her up to explore her abdomen, drain the area, and correct a ruptured organ if one exists. Follow-up care consists of 1 or 2 weeks of antibiotics. Give your cat antioxidant vitamins and minerals as advised by your veterinarian.

PREVENTION: Supervise your cat's playtime to prevent her from swallowing foreign objects or being, hit by a car or bike. Prevent your cat from getting into the garbage, which can contain fatty food that may stimulate pancreatitis (see section, Vomiting, Diarrhea, Excess Salivation, and Pain on the Right Side of the Abdomen Directly Behind the Rib Cage, pp. 132–133), which will add to your cat's discomfort. Spay your female cat so she cannot develop a pus-filled uterus (a condition known as pyometra) which can rupture into the abdomen.

135

CHAPTER 9

Spine, Limbs, and Paws

Cats are usually thought of as active, mischievous creatures—animals who are always ready for a game of chase-
the-ball, a fast climb up the drapes, or a quick scamper
through the house. Yet, in order to lead such an action-packed
life, your kitty needs strong, resilient muscles for smooth
movement and hardy bones to support her weight, cushion
the impact of all this verve, and protect her innards.

Though a number of ailments can affect the musculoskeletal system—from cancer to bone fractures to muscle sprains—
most produce locomotive symptoms. Such signals are usually
easy to pick up on. After all, when a cat can't—or won't—
move, or begins carrying herself in an unusual way, her loving
owner can't help but notice.

Change in Posture

Many things can cause a change in your pet's posture: An energy-
sapping illness like pneumonia will make it hard for a weakened cat to
stand up straight; physical trauma can alter a kitty's carriage depending
on what hurts; any ailment affecting an abdominal organ can leave an
animal hunched up in pain. Noting the symptoms that accompany the
change in posture is important.

Rigid or Flaccid Paralysis of the Front Legs or the Hindquarters, Cries of Pain When Picked Up or Petted Along the Spine, and Avoidance of Stairs and Jumping

RELATED SYMPTOMS: Your cat hesitates to play and may walk with a
stiff gait—or he may not be able to walk at all. He may lose control
of his bladder and bowels and may suffer from total paralysis of the
legs and tail.

POSSIBLE CAUSE: Is your cat older than five years? He may have a **herniated disc**, also known as **intervertebral disc disease**. This is a somewhat uncommon traumatic or hereditary condition wherein the outside covering of the disc becomes weak and eventually ruptures. The gelatin-like core of the disc leaks out and moves into spots where it puts painful pressure on the spinal cord or spinal nerve roots. It should be noted that this condition can affect discs anywhere along the spine.

CARE: **Take your kitty to the vet immediately.** If possible, carry him in a pet carrier or other container with a hard, flat surface so that his spine isn't subject to movement. If the cat can walk and the condition has stayed constant for several days, the circumstances aren't as dire. In either case, the vet will perform a physical exam and take radiographs to determine the location of the affected disc(s) or to distinguish disc disease from a fractured vertebrae.

Milder cases usually require cortisone, muscle relaxants, and physiotherapy, which consists of whirlpool baths, passive joint exercise, and the application of heat (via a hot-water bottle or heat lamp) to the affected region. A heating pad applied to the most sensitive area of the back 3 times a day can make your cat more comfortable. Since pain is nature's way of preventing your kitty from overworking a damaged joint (thus possibly injuring it further), many vets are hesitant to administer pain medication. In severe cases, surgery will be performed to remove the material that is pressuring the spinal nerve root or spinal cord. Acupuncture also has proved very effective in both reducing pain and relieving pressure on the spinal roots, consequently restoring use of the limbs.

Whether your cat has undergone surgery or more conservative medical care, certain homecare measures can make your pet more comfortable. Feed him a concentrated proteolytic plant enzyme (available at health-food stores) 3 or 4 times a day, and give him vitamin E. Vitamin C, chondroitin sulfate, and glucosamine sulfate can help strengthen the connective tissue surrounding the disc and vertebral joint (see Appendix E: List of Recommended Dosages, pp. 169–189). If you have stairs, don't allow your cat to use them: Carry the kitty up and down steps when necessary. Likewise, don't allow your pet to jump on or off anything elevated, including furniture. If you are used to active play and/or roughhousing with your cat, stop this interaction until the cat is well. Actually, the equivalent to human bed rest is the best advice for a cat suffering from a herniated disc.

PREVENTION: Although intervertebral disc disease cannot be totally prevented, keeping your cat nutritionally fit and at a healthy weight for his frame will make him more comfortable and less likely to develop a herniated disc. If he is already overweight, put him on a weight-reduction plan. Continual supplementation of his diet with the antioxidant vitamin C and with the cartilage and connective-tissue nourishing agents glucosamine and chondroitin sulfate can help minimize chances of further disc herniation (see Appendix E: List of Recommended Dosages, pp. 169–189).

Lameness

In vet-speak, **lameness** refers to any leg (or legs) that a cat is unable or unwilling to use in a normal fashion. With mild lameness, the gait may look almost normal, but on close inspection, the cat may "favor" the limb by putting less weight on it or bearing weight on it for a shorter time. With severe lameness, the cat will either hold his leg off the ground or literally drag the limb on the ground. For information about possible emergencies involving lameness, turn to the section in Chapter 1, Break, Fracture, or Sprain: Difficulty Moving, pp. 5–6. Back injuries or spinal-cord disease can also result in an abnormal gait.

The reason a cat can't or won't put his weight on a limb isn't always a serious one. In fact, should you notice your cat suddenly avoiding using a leg, check the cat's connecting paw. Quite often a **sharp object is stuck in the footpad** or **between the pads,** making it uncomfortable for the cat to bear weight on the leg. Other reasons for not using a foot include **puncture wounds, cuts,** or **abrasions.** A **cracked toenail** may also be a cause. As minor as it sounds, a broken claw can cause your cat much pain. A **sprain** or **strain to the leg** itself—due to overexertion or an accident—also keeps a cat off his limb. **Torn ligaments,** especially of the knee joint, are another common cause of lameness. Bruising and/or inflammation of muscles and tendons may also produce significant lameness.

If the lameness has come on more slowly, a more serious underlying medical cause, such as a **fracture** or **bone cancer,** may be to blame. In a young kitten, lameness can indicate that some of the limb **bones are growing irregularly,** whereas in an older cat, lameness can be a result of

arthritis. Lameness may also result from an injury to the muscles of the back, the spinal cord, or to the nerves leading to the leg muscles.

Lameness in One Leg, Swelling and Obvious Painfulness, and Noticeable Warmth in a Localized Spot of Leg Bone

RELATED SYMPTOMS: The cat will have a fever and the affected area may also develop a deep, oozing wound.

POSSIBLE CAUSE: Was your pet recently bitten by another cat or wild animal? Has he undergone any type of bone or soft-tissue surgery lately? Did he recently recover from an infection of an internal organ, the blood, or the skin? A "yes" to any of these may indicate a **bone infection**, known medically as **osteomyelitis**. Although animal bites and contaminated surgery are often to blame, an infection in another area of the body also can spread to the bone. Regardless of the cause, a bone infection can work its way through the skin, resulting in deep, draining wounds.

CARE: Take your kitty to the vet, who will perform a physical examination and take an X ray. A bacterial culture of any pus or liquid seeping from the site helps identify the bacterial or fungal organism responsible for the infection, thus making it easier to select an effective antibiotic. In addition to antibiotic therapy, surgery is sometimes necessary to establish better drainage and to remove bony fragments and debris. You can also feed your cat high levels of proteolytic enzymes and dimethylglycine (both found in health-food stores; see Appendix E: List of Recommended Dosages, pp. 169–189) to help support the immune system, remove pus and debris, and destroy the offending organism.

PREVENTION: Supervise your cat's play and attend to any wounds immediately.

Lameness in One or Both Rear Legs and Difficulty in Getting Up and Lying Down

RELATED SYMPTOMS: The cat has a swaying or waddling walk and may appear knock-kneed. Due to his difficulty in moving, the animal may appear lazy.

POSSIBLE CAUSE: Is your cat a Devon Rex, Maine coon, or Norwegian Forest cat? He may have **hip dysplasia**, a rare, partially hereditary disease of the hip joint that can affect certain breeds of cats. The condition begins as a cat is growing, caused when the hip joint develops improperly, and results in a loose-fitting and malformed ball-and-socket joint. Hip dysplasia is aggravated by excessive use of the joint, and it eventually develops into arthritis.

CARE: Hip dysplasia is not curable. If your kitty is an adult, not suffering severe pain, and not experiencing a worsening of the condition, you can take measures at home to make your pet more comfortable: Keep the environment warm and dry, don't let your cat jump or exercise heavily, and don't let him become overweight (extra weight stresses the hip joints).

As for diagnosing hip dysplasia, visit your vet. An X ray is required to make a definite diagnosis. A kitten who is still growing definitely should be seen by a vet. In severely lame cats, you may consider joint surgery. Nonsurgical options include giving your cat painkillers whenever his pain becomes severe. Ask your vet about acupuncture and gold bead implantation to relieve discomfort. The combined use of nutritional antioxidant supplements and glucosamine sulfate and chondroitin sulfate are very helpful in treating the condition and reducing joint pain (see Appendix E: List of Recommended Dosages, pp. 169–189).

PREVENTION: If your kitten is one of the aforementioned breeds, have him X-rayed for the detection of hip dysplasia, thus allowing early treatment if needed. For large cats of all ages, conscientiously maintain normal body weight. Although this will not prevent hip dysplasia, it may prevent a mild case from becoming severe.

Lameness in One or More Legs and Broken Nails on Affected Paws

RELATED SYMPTOM: The broken toenails will still be partially connected to the paw.

POSSIBLE CAUSE: Do you clip your cat's toenails infrequently, if at all? Has your cat been in a fight recently? There's a chance **one or more of his nails are broken**—an uncomfortable situation in which a claw hangs by just a thread of connective nail. This is one of those minor conditions that can cause pain severe enough to keep a cat off his affected foot.

CARE: If there are no infected areas (i.e., the adjoining toe is not severely swollen), you can try to take care of this yourself. The condition can be very painful if the nerve is exposed, so, before proceeding, you may want to wrap the cat in a towel to keep him from scratching you. It's helpful to have someone hold the cat, petting him and keeping him calm (and distracted). If the nail is hanging by a thread, take a pair of needle-nosed pliers or tweezers (or you can use your fingers) and give a quick, hard tug and twist at the end of the cracked, broken toenail. If the broken nail is attached more firmly, a nail cutter may be a better instrument to use. Your kitty probably won't like this, but the broken end needs to come off. Light bleeding can be stopped by placing a bandage against the area and applying pressure for 3 to 5 minutes. If the broken nail is cracked but still quite closely attached, I recommend you visit your vet: Cutting off the broken end of such a nail fracture can cause a good bit of bleeding.

PREVENTION: Because overgrown toenails are most likely to break, keep your cat's nails trimmed. (See Appendix A, Checklist for Good Health, p. 156).

Limping on One Leg and Licking the Connecting Paw

RELATED SYMPTOMS: You may notice small smears of blood on the floor where your cat has set down the affected paw.

POSSIBLE CAUSE: Could your cat have stepped on a sharp piece of glass, a thorn, or other sharp material during an unsupervised moment? He may have a **cut** or **puncture wound on his footpad.** This is easily determined by looking carefully at the bottom of your cat's foot and between the toe and pads.

CARE: If the cut is a deep one, your goal is to prevent severe blood loss. Immediately press a towel or bandage to the area and apply pressure until the bleeding stops. If necessary, you also can apply a tourniquet in addition to the direct pressure. Tourniquets are only used on appendages—the limbs and tail. To apply a tourniquet, find a soft, elastic fabric, such as a sock. Tightly tie the fabric around the appendage, directly *above* the wound. To avoid killing living skin and muscle tissue, you will have to loosen the tourniquet every 10 minutes, for 30 to 45 seconds at a time, to allow blood to flow into the limb. Once the blood flow has significantly slowed, apply a pressure bandage. A cut that requires pressure to stop the blood flow often requires stitches, so continue to hold the compress in place as you take your pet to the vet (a second person could help you with this).

Always examine the injured area for foreign material. If the footpad is free of foreign debris, wash it with 3% hydrogen peroxide, soap, and water, then rinse and dry thoroughly. Apply an antibiotic ointment (such as Neosporin), and wrap the paw with a gauze bandage to keep it clean and to help speed healing.

If something is embedded in the flesh of the paw, you can try to remove it. Start by wrapping your cat in a towel so that his legs—and claws—are prevented from moving. Even the gentlest feline can get scared enough to bite or claw at you. After all, he doesn't know what you're doing with those tweezers—only that you're messing with a foot that already hurts! If the object can be easily removed with tweezers, quickly yank it out. If it is embedded too deeply, you may need to coax the object to the surface with a *sterilized* sewing needle or straight pin, much like you would when removing an embedded splinter in your own hand. Once the intruder is at the surface, you can use your tweezers to pull it out. Finish by washing the area with soap and water and/or 3% hydrogen peroxide, then drying it, and wrapping a gauze bandage around the paw. Wrapping adhesive tape over the gauze and attaching it to a small amount of hair may be necessary to keep the bandage on. Any bleeding should stop within 5 minutes, and the cat should be putting pressure on the foot within 2 to 3 hours.

If you haven't removed the entire object, if the animal doesn't put pressure on his foot after 3 hours, or if the area becomes more red and swollen within the next 2 days, take your kitty to the vet.

PREVENTION: Supervise your cat's playtime and attempt to keep dangerous objects off the ground and away from your cat's exercise area.

Limping on One or More Legs, with a Raw-Looking Footpad on Affected Paw

RELATED SYMPTOMS: When you examine the footpad of the leg your cat is limping on, you may also notice blisters, cuts, or flaps of hanging skin. The cat may lick the affected paw(s).

POSSIBLE CAUSE: Has your cat been walking on salt-treated surfaces (during the winter) or hot asphalt (during the summer)? Has your pet been traversing rough, rocky terrain? A "yes" to either of these may point to **footpad abrasion**. The condition is caused when the tough, protective skin of the footpad is worn away, exposing the tender tissue beneath.

CARE: If infection has set in (indicated by the presence of oozing pus), **take your cat to the vet**. Otherwise, if the abrasions are superficial and not too severe, you can treat this at home. Wash the area with 3%

hydrogen peroxide, warm water, and mild soap, rinse, then thoroughly dry it. Next, swab the footpad with an antiseptic solution, such as chlorhexidine. An antibiotic ointment, such as Neosporin, can be applied to particularly raw spots to ward off infection. You can attempt to seal a superficial wound abrasion (thus reducing pain and protecting the wound from further contamination) by using a product such as New Skin, or tincture of benzoin (but be aware that these substances may momentarily sting your pet). Bandage the paw to keep it clean and to prevent your cat's rough tongue from further irritating the wound.

It is next to impossible to keep your cat off his paws during the healing process, which may take up to 2 weeks. If your kitty refuses to carry the paw or paws, heavily bandage the paw(s) to keep them from directly contacting the ground.

PREVENTION: Avoid letting your cat walk on hot asphalt or salted sidewalks. Should your kitty's paws come in contact with salt (which dries skin and causes it to crack), manually remove the salt granules and wash the paws with water and mild soap. Rinse well and dry thoroughly.

Mild to Severe Lameness in Hind and/or Front Legs Upon Standing, Thickening of the Joints of Affected Legs, Overall Stiffness, and Apathy

RELATED SYMPTOMS: Although it doesn't disappear, the lameness usually improves after the cat has had mild exercise. If you gently move the affected joints, you may hear a grinding noise of cartilage or bone grating against cartilage and/or bone. You may notice the presence of a mild fever that seems to randomly appear, disappear, and reappear. These symptoms may have developed slowly over the course of several years.

POSSIBLE CAUSE: Has your cat experienced a joint injury or joint infection at any time in the past? Is he an older animal? He may suffer from a type of **arthritis**. The word "arthritis" simply means joint inflammation. The disease has many types and causes, including old age and previous trauma to a joint due to infection or injury. In the case of congenital arthritis, some breeds are genetically predisposed to getting arthritis.

CARE: If you're unsure your kitty has arthritis—or if your cat's lameness is severe—visit your vet, who will perform a physical exam and take X rays. In moderate to moderately severe cases, your vet may prescribe anti-inflammatory pain medication and nutritional supplements, such as antioxidants, glucosamine sulfate, and chondroitin sulfate, which suppress inflammation, help squelch the body's destruction of its own joint structures, and help repair damaged cartilage. In cats predisposed toward arthritis, these remedies may need to be continued for life.

In more severe cases, loose cartilage may be present in the joint. If so, your veterinarian will surgically remove the detached matter. If the arthritis is mild, there are easy homecare measures you can employ that will make the cat more comfortable. First, since extra weight taxes the joints, an arthritic, overweight cat must thin down. Also, keep his environment warm: Cold air makes joints stiff and achy.

Moderate exercise—and the operative word here is "moderate"—is an essential part of homecare. (No rough or extended games of cat-and-mouse and no jumping from high places.) Instead of allowing your cat to sleep on hard surfaces, provide soft bedding for comfort's sake.

PREVENTION: You can't prevent arthritis in a cat genetically inclined to the condition, but you can ward off the ailment in a nondisposed animal by helping him avoid joint injuries, such as those a cat may get when hit by a car. Keeping your cat at an ideal weight will take stress off his joints and may help prevent the onset of arthritis.

Sudden Lameness and Collapse of Rear Legs

RELATED SYMPTOMS: One leg might be more severely affected than the other. The muscles may be swollen, and your cat may be in extreme pain.

POSSIBLE CAUSE: Has your cat been diagnosed with a heart condition—or is he an older cat who could be at risk of heart disease? He may have **thromboembolism of the iliac arteries.** The iliac arteries supply your cat's back legs with blood. In the presence of a heart disease (such as cardiomyopathy; see section in Chapter 7, Difficulty Breathing, Possibly Accompanied by Lameness or Paralysis, Loss of Appetite, and Lethargy, pp. 76–77), a blood clot can develop, blocking the artery that supplies the blood to the hind legs. The lameness can be blamed on the resulting lack of circulation to the muscles and nerves.

CARE: Take your cat immediately to the vet, who will usually be able to diagnose the condition by a physical exam. In cases where heart disease is mild (meaning easily treatable) and the blockage is partial, there's a chance your cat will regain use of his legs. When heart disease is severe, however, the chances that blood supply can be restored—thus restoring your cat's use of his limbs—are not good, and 60 to 70 percent of cats with thromboembolism of the iliac arteries die within days of developing the condition. Those animals who live are treated for congestive heart failure and prescribed medication to prevent future clots.

PREVENTION: If your cat has heart disease—or is an older cat who is at risk for heart disease—do everything you can to treat it and minimize the risk of clot formation. Although use of aspirin is not normally recommended for cats, it can aid in the prevention of repeated arterial obstruction for cats who have already suffered one blood clot. But follow the advise of your veterinarian before considering giving your cat aspirin.

Swelling and Tenderness in a Leg or the Spine

RELATED SYMPTOMS: Your cat may lose interest in food. He may be reluctant to move the affected body part and may shy away from being touched there. If a leg is affected, he may have a mild limp that can become progressively worse.

POSSIBLE CAUSES: Is yours a mature cat? He may have a **bone tumor.** The good news is that bone cancer is relatively rare in cats. The bad news is, the majority of skeletal tumors are malignant.

CARE: When malignant, skeletal cancers spread quickly, usually to the lungs. Therefore, take your cat immediately to the vet, who will perform a radiographic exam to determine the presence of a tumor. A biopsy of the tumor can determine whether it is benign or malignant.

If the growth is malignant, your vet will take chest X rays to see if the cancer has spread to the lungs. If the lungs have not been affected, he may recommend removal of a portion of the limb. If two or more tumors are found on a leg, the entire limb may be amputated. Though this sounds extreme, most cats adapt well to three-legged life. To prevent the cancer from spreading to another body site, chemotherapy often follows tumor removal and/or amputation.

PREVENTION: Nothing can be done to prevent bone cancer, other than maintaining a healthy body, which in turn will keep the immune system strong.

Stiff or Unusual Movements

Falling over, trembling, stiff legs, a disjointed gait—all are admittedly scary signs. Often they appear with such ailments as distemper, heart disease, tumors, or poisoning, and may be accompanied by other symptoms, such as coughing, vomiting, or fever. When symptoms of incoordination come on quickly, you are wise to visit your vet immediately.

Skipping and Frequent Lifting of One or Both Hind Legs

RELATED SYMPTOMS: If both legs are affected, the cat will hop with his back rounded like a rabbit. You may notice thickening of the affected knee joint(s).

POSSIBLE CAUSE: Is yours a Bengal cat, British Shorthair, Chartreux, Devon Rex, Egyptian Mau, or Turkish Angora? Has your cat been in an accident at some point in his life? He may be suffering from a **slipping** or **dislocated kneecap.** This condition, also called **luxating patella,** is congenital in many breeds or can be blamed on past physical trauma. Due to the malformation of the bones forming the knee joint, the kneecap slips inward out of its normal position in the joint. Depending on the severity of the deformity, the kneecap may slip in and out of place intermittently or move out of place permanently.

CARE: Very mild cases cause only intermittent lameness and do not need treatment, although you may want to take your cat to the vet for a professional diagnosis. In more severe cases, involving serious lameness, your vet can surgically move the kneecap back into place and stabilize it, allowing your cat near-normal movement. To help strengthen the connective tissue surrounding the knee joint and repair injury to the cartilage of the knee, give your cat vitamin C, chondroitin sulfate, and glucosamine sulfate (see Appendix E: List of Recommended Dosages, pp. 169–189).

PREVENTION: There is no prevention. If you plan to adopt a cat who is one of the breeds just listed, ask a vet to help you choose the individual animal. Having a vet inspect an animal before you decide to keep him can help ensure that you don't unknowingly select a kitty with this problem. To avoid "creating" kittens with this condition, cats with slipping kneecaps should not be allowed to breed.

CHAPTER 10

Tail and Anus

When most of us think of our cat's back end—if we think of it at all—the tail is probably the first thing that comes to mind. Yet, that's not all there is to your kitty's posterior. It's wise to pay attention to the anus, too—it's an excellent harbinger of your kitty's overall wellness.

Yes, this particular body part has an unsavory reputation, which is unfortunate. The anus works hard to finish off the digestive process started by the mouth and stomach. A change in anything leaving the anus can indicate a small overall health problem that can be caught and addressed while still minor. Changes in feces can also hint at a more major ailment involving the intestines, prostate, or other internal area. Of course, the backside can be plagued by a few illnesses that strike solely this area, such as prolapsed rectum or blocked anal glands. But if you keep a careful watch on your cat's behavior and appearance, you'll know when there's a problem.

Conditions of the Anal Region

You may notice something is wrong with an unfamiliar part of your kitty's backside: the anus. If you catch your cat repeatedly sniffing or licking the area or dragging it along the ground, you can be pretty sure that something back there is bothering your pet. This something could have a nonmedical cause, such as your cat sitting on something irritating or sticky, not cleaning herself thoroughly after eliminating, or not getting all the soap rinsed off after a bath. Since longhaired cats are especially susceptible to these irritating situations, keep the fur around the backside clipped short. This will prevent anal secretions and bits of feces from becoming embedded, which smells bad and also may make your kitty itch. Unfortunately, such behavior also can indicate a health condition that directly targets the anus and anal glands, such as **an abscess or tumor.**

Puckered Skin Protruding from the Anus

RELATED SYMPTOMS: Your cat may lick the area and you may notice straining during defecation.

POSSIBLE CAUSE: Does your cat eat chicken bones? Does she drink water infrequently? Is she frequently constipated? Has she been diagnosed with hairballs? She may have a **prolapsed rectum.** The protruding, puckered skin you see is actually an interior portion of the rectum that has been forced out through the anus, caused by straining to pass hard feces. These hard feces can be blamed on anything from consumption of chicken bones to lack of moisture in the diet to hairballs.

CARE: There are several things you can do at home to relieve your pet's constipation. Try replacing half of your cat's food with freshly ground, high-quality raw meat (a natural laxative) or add powdered psyllium, such as Metamucil (see Appendix E: List of Recommended Dosages, pp. 169–189) to her diet to keep her stool soft. You also can mix mineral oil in the cat's food 2 times a day, for no more than 1 week, to relieve constipation. If the cat is still suffering from constipation and persistent rectal prolapse, visit your vet, who will medicate the tissue, lubricate it with petroleum jelly or K-Y lubricating jelly, and replace the rectum manually. To prevent recurrence, your vet may place a temporary "purse-string" suture around the anus. This will hold the rectum inside the anus where it belongs until the area is healed. Further veterinary treatment includes recommending an easily digestible cat food and a feline laxative. When your cat begins voiding in a more normal fashion, the rectum will have much less of a tendency to prolapse.

PREVENTION: To prevent constipation, which, in turn, can lead to straining and rectal prolapse, encourage your kitty to drink generous amounts of water. Don't feed your cat feces-compacting bones. To keep hairballs from forming inside your kitty, groom your pet weekly and feed her a hairball preventive such as Laxatone.

Swollen Areas or Abscesses Around the Anus, Scooting the Anus Across the Ground, or Constant Licking of the Anus

RELATED SYMPTOMS: The area will often smell foul. Swelling around the anus may be the only symptom, most noticeable at the "4 o'clock" and "8 o'clock" positions around the anus. Abscesses of the anal glands may break open to reveal blood and pus, leaving a craterlike wound.

POSSIBLE CAUSES: Is yours a mature male cat? Is he an unneutered cat? A "yes" to either of these questions might indicate a condition concerning the anal glands, specifically **clogged anal glands** and **anal-gland abscesses**. (Although female cats also have anal glands, the conditions primarily affect males.)

To understand anal disorders, an anatomy lesson is in order: Anal glands, also called anal sacs and often referred to as "skunk glands," sit at the "4 o'clock" and "8 o'clock" positions around the anus (when you imagine the anus as the center of a clock). They are lined with cells that manufacture an extremely foul-smelling liquid. As feces make their journey out of the anus, pressure is placed on these sacs, and they empty their stored material through tiny ducts located alongside the anal opening. It's speculated that these sacs and their smelly secretions help cats mark territory and identify each other, though no one has ever proven this.

If these anal sacs aren't emptied on a regular basis, the long-standing secretions thicken into a pasty, gritty sludge. Instead of passing easily through the tiny duct, this pasty ooze gets stuck in the anal sacs—much like the problem of trying to squeeze toothpaste through a pinhole. To make an ugly situation even uglier, bacteria sometimes work their way into the filled-up sacs, causing the clogged anal glands to become infected and abscessed.

CARE: Take your cat to the vet, who will examine the area to determine the problem. Clogged anal glands can often be emptied by manually squeezing out the contents, but this should be left to a veterinary professional! If done incorrectly, the procedure can drive the material deeper into the anal sacs or injure the glands. If these sacs cannot be manually emptied, they may need to be flushed out using a syringe and special needle.

Abscessed anal glands are lanced, irrigated of pus and blood, swabbed with a diluted antiseptic solution, and packed with an antibiotic ointment. If the abscesses recur at a later date, the glands may be surgically removed.

PREVENTION: Feeding your pet high-fiber cat food adds bulk to his diet, promoting larger stools and, therefore, a more thorough compression and emptying of the sacs with each bowel movement. This lowers the recurrence of clogged anal glands and abscesses. To keep a clogged anal gland from becoming abscessed, take your kitty to the vet as soon as the animal begins scooting and/or licking at the anal area.

Conditions of the Tail

Wondering what kind of mood your cat is in? Look at her tail: High in the air and she's proud and ready to take on the world; puffed like a bottlebrush and something has her on edge; swishing gently back and forth and she's ready to hunt. Some owners place so much importance on a cat's tail movements that to see the appendage stuck in one position can be scary. Yet injury to one or more of the tail's vertebrae or to the region where the tail joins the pelvis can do just that, freezing the tail in a limp position between the legs.

Bent-Looking or Limp Tail That Doesn't Move

RELATED SYMPTOMS: The cat may yelp when her tail is touched or may shrink from being pet on her backside.

POSSIBLE CAUSE: Does your cat play outside unsupervised? Was she recently in an accident or in a fight with another cat? Did she have her tail slammed in a door? Your kitty's **tail may be fractured, dislocated, or bruised.** An injury to the nerves that lead to the muscles of the tail can produce paralysis in that area.

CARE: Take your pet to the vet. Because animals suffering from broken bones often scratch or bite out of pain and fear, wrap your cat in a towel to keep her from wounding you, your vet, or your vet's assistant. An X ray can confirm whether any bones have been fractured or dislocated. If so, the tail will be set and splinted.

A wound, either from a bite or cut, will be dressed or stitched if necessary. Should your kitty suffer from a severe bruise, your vet may suggest that you keep the animal calm and comfortable indoors for 2 weeks so that the hurt can heal. Injuries causing a tail paralysis will hopefully heal with time. If not, the tail may require amputation.

PREVENTION: Supervise your cat's play.

Wound on the Tip of the Tail, Accompanied by Bleeding

RELATED SYMPTOMS: The wound doesn't heal, even when you try to bandage the area. The skin may be abraded, and you may be able to see underlying muscle and/or bone.

POSSIBLE CAUSE: Has someone accidentally slammed your kitty's tail in a door or stepped on it? Could your cat have caught her tail in a gate, lawnmower, or on some other sharp object during an unsupervised moment? The animal may have **injured the tip of her tail**. Unfortunately this is one body part that is relatively slow to heal and may be easily reinjured.

CARE: Take your cat to the vet. Anytime your cat moves her injured tail, she risks hitting the appendage on a nearby wall, the floor, or your leg. Wrapping the tip in a heavy layer of bandaging shields the wound, but if your cat is like most kitties, she probably will try to chew off this bandage. Your vet will probably recommend using an Elizabethan collar, making it harder for your pet to reach around and nip her own tail. (However, because cats are so flexible, they may still be able to reach their tails.)

PREVENTION: Watch out for your pet's tail when shutting house and car doors.

APPENDIX A

Checklist for Good Health

Good illness-preventing homecare is the key to a healthy, happy cat. Fortunately, basic health maintenance is easy. In fact, it's much the same for cats as it is for humans: Provide healthy food and fresh, clean water, regular exercise, proper grooming, even check regularly for signs of disease so that any condition can be promptly addressed and managed before it grows more serious. Here's a rundown on how to look after your pet so the two of you can enjoy a healthy, happy life together.

Normal Vital Signs

A cat's vital signs provide an important glimpse into the state of his health. If your pet's vital signs differ from any of the following, there may be an underlying medical reason. Consult your vet. (See Appendix B, pp. 160–162, How to Perform a Weekly Home Exam, for information on gathering this information yourself. See the Symptoms Quickfinder for ways to recognize signs of illness.)

Temperature: 100° to 103°F (the average is 120).
Resting heart rate: 110 to 140 beats per minute (the average is 120).
Resting respiration: 20 to 30 breaths per minute.
Weight: Unneutered males typically weigh 8 to 15 pounds. Females and neutered males generally weigh between 6 pounds and 12 pounds.

Picking Up and Holding Your Cat

Some cats adore being held; quite a few others do not. Whichever camp your kitty falls into, handle him properly—both to avoid injury and to make the experience more pleasant for both of you. When picking up your cat, try to position yourself directly behind him. Gently place your dominant hand under his chest and begin lifting, quickly positioning your other hand under your kitty's hindquarters. Let his sternum (chest) rest in your loosely cupped hand as you bring him to your chest.

When holding your cat, you have two choices: Which you use depends greatly on which your kitty is most comfortable with. Position one entails your pet relaxing chest-level in your arms with his paws resting in the crook of your arm. Position two lets your kitty sit in a more upright pose, with his paws on your shoulder and his hindquarters supported by your arm. If yours is a small kitten, you should hold him *softly* with both hands.

Cats, being the independent thinkers they are, want to control how long they're held. They'll start squirming to let you know when they want to get down. As soon as you feel your pet getting restless, put him down immediately. If you ignore his requests, he may resort to nipping or scratching you in order to make his point.

How Curiosity Can Kill the Cat

Live with a cat for any amount of time (even just a few hours)—and you'll see just how inquisitive these creatures are. Although this detective instinct helps felines in the wild, it can harm them in your home. Unlike dogs—who can't jump to a high shelf to nose around a particular houseplant and can't climb a shower curtain to reach a roll of dental floss—cats are unbelievably mobile. From the moment you bring your feline friend home, be prepared to keep anything that represents a health hazard secured in a shut cupboard, closet, dresser, or trunk. Countless owners, however, tell tales of housecats who learn how to open cupboard and closet doors. Therefore, make sure that your stashing spot is absolutely paw-proof. Among the items you'll want to keep out of reach are: sharp fasteners like tacks, staples, and nails; cosmetics and toiletries; glues; cleaning products; any threadlike product; breakables; matches; insecticides; exposed wires; and any plants your vet has labeled as hazardous.

Preventive Health-Maintenance Schedule

- Heartworm test: Yearly.
- Neutering (castrating/spaying): 3 to 6 months of age.
- Feline distemper vaccine: Yearly booster*.
- Rhinotracheitis-calici vaccine: Yearly booster*.
- Feline leukemia test: Given prior to first feline leukemia vaccination.
- Feline leukemia vaccine: Yearly booster*.
- Rabbies vaccine: 1- or 3-year boosters, depending on the incidence of rabies in the area.
- Routine stool check: Twice yearly.
- Dental exam and teeth cleaning: Once to twice yearly.
- Routine physical exam: Once to twice yearly.
- Routine blood screening: Yearly.

*A growing number of veterinarians are questioning the need for life-long yearly boostering. Instead, many recommend that the cat's blood be tested yearly to determine if protective levels of antibodies are present. If so, no vaccines are given. If antibody levels are low, a booster is given.

Home Emergency Kit

Should an emergency ever arise, you'll save precious time with a preassembled kit of emergency hardware. House the following essentials in a small cardboard box, fishing tackle box, or Tupperware container, and keep the kit in a convenient (yet pet-safe) location:

- Baby oil.
- Bulb syringe.
- Charcoal suspension.
- Chlorhexidine solution (an antiseptic).
- Cotton balls.
- Cotton swabs.
- Dramamine (over-the-counter; for motion sickness).
- Ear wash and wax solvent solution.
- Gauze pads (3 inch × 3 inch).
- Gauze roll (3 inch).
- Hydrogen peroxide solution (3%).

- Iodine.
- Kaopectate.
- Milk of magnesia.
- Neosporin ointment.
- 1-inch roll of adhesive tape.
- Pepto-Bismol.
- Rectal thermometer (preferably electronic).
- Rubbing alcohol.
- Scissors.
- Small jar of petroleum jelly.
- Styptic pencil.
- Tweezers.
- Wound powder.

Good Grooming

Cats do a thorough job of keeping themselves clean. Still, a little homecare can't hurt things: In fact, if you have a mediumhaired or long-haired cat, you *must* brush him often to get rid of matted fur and debris that he can't remove himself. Regular grooming also makes good health sense: By keeping hair trimmed around ears, eyes, and anus (if needed), eradicating snarled, matted fur, and clipping nails, you make it harder for your cat to suffer irritation and infection.

Another benefit of good grooming is that health problems, such as parasites or a skin condition, are more quickly (and easily) noticed when you regularly groom your cat. Here are a few good-grooming pointers:

EYES: Keep hair out of the eyes to prevent them from becoming irritated and infected. If your cat has long facial fur that grows into or near the eyes, be vigilant about keeping this hair clipped and out of the eyes. Any secretion buildup can be gently removed with a soft cloth or cotton ball. To clear away hardened secretions, dampen a tissue or cloth with warm chamomile tea or water to soften the deposit and make removal easier.

EARS: Is there a buildup of earwax or dirt lining your cat's ear flap? If so, remove it. Moisten a soft cloth or tissue with baby oil and gently swipe the surface. Because dirt and wax also settle on hair that happens to be growing in the ear, you should remove strands by simply pulling them out with your fingers. Avoid trimming these hairs with scissors: You run the very real risk of shredding your pet's ear. An earwax solvent should be placed in the cat's ear weekly to help remove wax and dirt that can settle in the bottom of the ear canal.

TEETH: If your cat has a dental problem, such as excessive tartar, your vet may ask you to brush the animal's teeth daily or every other day. Fortunately, this isn't difficult. Feline toothpastes are generally pleasant tasting—at least most cats think so!—and are available from your vet or pet store, as are various types of brushes. Apply the paste to the brush and employ the same movements you use for your own teeth: small, massaging circles at the outer and inner gumlines. Other dental-care products exist that only require application to the gums and eliminate the need for brushing.

NAILS: Too-long nails can break off or tear at the quick, causing the cat enough pain that he may refuse to walk. Thus, it's important to check your kitty's nails during each home exam. To trim a cat's claws, hold the paw horizontally. If you lightly press the footpad under the "toe," a nail should emerge. Does the tip of the nail curve below an imaginary straight line emanating from the bottom of the paw? If so, *clip just that part that curves under*—no more, since you risk cutting into the very sensitive portion of the nail that contains nerves and blood vessels. For best—and easiest—results, use a pair of cat clippers, which are available at your pet store.

COAT: How you care for your pet's coat depends greatly on what type of fur he has. Longhaired felines should be brushed at least every other day: Begin by detangling hair with a wide-toothed metal comb, then finish by stroking a wire brush through the hair. For cats with shorter fur, you can skip the detangling step and stick to once-a-week brushing.

Cats do such a good job of keeping themselves clean that baths are usually unnecessary. That said, your vet may advise a soak in the tub if your cat gets into something and obviously needs to be washed down. Should this happen, use warm water and an extremely mild pet shampoo, available at a pet store or from your vet. Do not use dishwashing soap: It is too harsh and can irritate and dry the skin. After wetting the animal's fur, pour shampoo into your hand and lather all body parts, being careful to shield the eyes. (Be warned: For your skin's sake, you may want to wear vinyl gloves. Your cat will probably struggle and may claw at you in an attempt to escape—most cats hate being wet!) Rinse your cat very thoroughly with clean, warm water. Towel-dry the kitty and either blow-dry fur (if he'll let you) until completely dry, or keep the animal in a warm, draft-free room until his fur dries.

ANUS/GENITAL REGION: Again, cats being the meticulous cleaners they are, few healthy felines let their rear ends become dirty. Should you notice bits of feces stuck to the fur or the hair soaked with urine, immediately clean the area. (You may want to use hypoallergenic baby wipes.) If your pet has long hair on his backside, you might consider clipping this fur short enough in order to keep it "out of the way."

Caring for a Sick Kitty

At some time in your cat's life, he may develop a condition that requires an operation or the administration of daily medication. Here's some advice designed to help you help your pet:

GIVING PILLS: Cats hate to swallow pills. Some pet owners have luck hiding medication in a wad of food, but others find that their cunning felines find a way to spit up the pill while swallowing the food. If your kitty falls into the latter category, you can dissolve the pill's contents in liquid (honey or liquid from tuna or sardines work well to mask the taste of the medicine). Your cat may lick it right off of your finger. If not, tilt your cat's head up. Using an eyedropper or syringe, place the liquid into the cheek pouch between the cheek and gums, which helps it to seep down the throat easily. Or, you can tilt the animal's head upward, hold open his mouth and, with your free hand, lay the pill as far back as you can on the tongue. Swiftly close his mouth and let the kitty swallow. Continue to hold the cat's head in an upward tilt: This further "encourages" swallowing. Once you're sure the pill has gone down the throat, you can let go.

TEMPERATURE-TAKING: Rare is the cat who will hold a thermometer under his tongue. This means that if you want to take your kitty's temperature, you must try another route. Enter: the rectal thermometer, an inexpensive, electronic thermometer that is safer than glass and is available at most drugstores. To use, apply a thin coating of greasy lubricant to the instrument. This can be petroleum jelly, vegetable oil, or a product such as K-Y lubricating jelly. Raise your pet's tail with one hand and insert the thermometer with the other, pushing firmly but gently with a subtle twisting motion. You may feel slight resistance—most likely, this means the thermometer is passing through a fecal mass. However, in the event that the thermometer could be caught on the side of the rectum, you may want to try redirecting or reinserting the device.

How far the tool needs to be inserted depends on the animal's size. An inch may offer an accurate reading for a small to medium cat. For a larger cat, however, you may need to insert the thermometer 1½ inches. The instrument should be left in for 3 minutes or less, depending on whether it is glass or electronic. If you use a glass thermometer, roll the thermometer between your fingers until you can see the thin line of mercury inside, or, in the case of the electronic thermometer, simply read the digital display.

MEDICATED BATHS: Skin conditions, such as eczema, and external parasites, such as fleas, ticks, and mites, may require that your cat be bathed with a medicinal shampoo. When using such a product, be vigilant about keeping the product out of your pet's sensitive eyes and ears, and do not let your pet swallow any of the shampoo-tainted water. After lathering the product into your cat's coat, allow it to set 5 to 10 minutes. Rinse with clean, warm water and towel the kitty dry. As for the required frequency of medicated baths, talk to your vet. Depending on the condition being treated, 1 to 2 times every week is the norm.

GIVING EYEDROPS OR APPLYING OINTMENT TO THE EYE: If your kitty is ever diagnosed with an eye condition, you'll probably have to administer eyedrops or ointment. To give your cat drops, stretch the bottom lid slightly away from the eye and squeeze 3 drops inside the lid. For ointment, pull up on the upper eyelid slightly and squeeze a ⅓-inch strip of ointment under the lid.

CARE OF WOUNDS AND SURGICAL INCISIONS: If the wound is bandaged, leave the area alone until your vet removes the wrapping. However, larger wounds are usually left uncovered in order to speed healing. If the wound is especially deep or infected, the vet may even insert a drain to allow fluids to escape. An uncovered incision should be kept clean by dabbing at the area 2 times a day with a clean, soft cloth or tissue. Do not apply an ointment or any kind of lotion unless directed.

INSULIN INJECTIONS: Should your cat ever develop diabetes, you'll have to give him daily insulin shots. (Your vet will give you the information you need to obtain syringes and insulin.) Though the thought is unpleasant, the actual task isn't difficult. Start by finding a loose fold of skin: The neck, back, and sides of the torso are ideal. Pull up a flap of the cat's hide into a kind of tent shape. Insert the needle into the bottom of this "tent," making sure the needle has not passed out the other side of the skin tent. To make sure that you are not in a blood vessel, always pull back on the barrel of the syringe before infecting. If blood is seen in the syringe when you draw back, remove the needle and insert it at a different site. Once you are certain that the needle is not in a blood vessel, inject the contents of the syringe. To prevent a buildup of scar tissue, you must inject the insulin in a different spot each time.

ENEMAS: Should you need to give your cat an enema, purchase a fleet saline or mineral oil enema from your local drugstore (cats should not receive any other type of enema). Place your cat in a bathtub or take him outside, then wrap him in a towel or get someone to help you restrain him (the cat will probably struggle a great deal). Be sure that his rectum and tail are exposed, then lubricate the tip of the enema bottle (if it is not a prelubricated type). Raise the cat's tail and insert the enema bottle approximately 1-inch deep into the rectum. Squeeze the contents into the cat's colon, then remove the bottle. Hold his tail down over his anus for several minutes (if possible) to keep the solution from coming out, then let your cat expel the liquid. If no bowel movement is produced, repeat.

How to Perform a Weekly Home Exam

Head and Neck

- **Symmetry**: Compare the features on the right side of the face with those on the left.
- **Eyes**: Check for clarity, pupil size, and excess discharge. Notice the color of the globe and inner surface of the eyelid.
- **Ears**: Learn the normal skin color. Note the odor of the ear and observe the amount of hair and wax in the canal.
- **Nose**: Check that the nostrils are open and look for any nasal discharge.
- **Mouth**: Check the color of the gums. Look for any sores and growths. Examine the teeth for tartar and look for any missing or broken teeth. Look under the tongue for growths, lacerations, or foreign bodies.
- **Trachea (windpipe)**: Learn its normal size, shape, and location.
- **Lymph nodes and salivary glands**: These are located below the ears, just behind the point where the lower jaw bends vertically. There are also lymph nodes located further down the neck, just in front of where the neck meets the body. Notice the normal size and shape of these glands so that swelling will be more readily detected.
- **Thyroid gland**: Run your thumb and forefinger down either side of the windpipe, starting from just below the "Adam's apple" or larynx and moving to the thoracic inlet. If the thyroid is enlarged, you may feel it pop under your finger. If it's normal, you won't feel anything unusual.

Trunk (Body Proper)

- **Symmetry**: Compare the features on the right side with the left. Observe the degree of prominence of the ribs, hipbones, and backbone.

- **Abdomen:** Compress it with one hand on each side and note any tenseness, tenderness, or distension.
- **Mammary glands:** Note the size of the nipples and glands. Feel for lumps (in both sexes).
- **Genitals:** Examine for any swelling of the prepuce, scrotum, or vulva. Note any odor or discharge. Look for redness, chapping, or irritation of skin surrounding scrotum, prepuce, or vulva.
- **Rectal area:** Note its color and appearance. Check for fleas, tapeworm segments, dried stool, rectal protrusion, growths, and swollen anal glands.
- **Lymph nodes:** These are located in the armpit region of the front legs and in the groin region of the back legs. Note any lumps in these areas.

Limbs

- **Symmetry:** Compare the bones and joints on the right with those on the left.
- **Gait:** Observe your pet walking and running from several angles, including the front, back, and side views.
- **Anatomy:** Note the angles and relationships the bones have with one another.
- **Range of motion:** Move each joint through its full range of normal motion. Note any grating or pain the cat experiences.

Skin, Hair, and Nails

- **Skin:** Part your cat's hair and look for flakes, pimples, scales, scabs, cuts, tumors, cysts, fleas and flea droppings, ticks, redness, and abrasions.
- **Hair:** Note its luster, texture, thickness, dryness, oiliness, and any areas of hair loss.
- **Nails:** Check their length and look for any split, broken, or ingrown nails.

Pulse

- It can be easily felt in the groin at the uppermost part of the inner thigh where the leg meets the body. Note its rate, strength, and rhythm.

Heart Rate

- Place your hand on your pet's chest and feel the beat. Note the rate and rhythm.

Respiration

- Become acquainted with your pet's normal rhythmic chest motion. Note its rate and rhythm.

Eating, Drinking, Urinating, and Defecating

- Pay attention to your pet's normal motions as he performs these natural functions. Note the color and consistency of the cat's stool and urine.

Posture

- Observe from several angles and note the carriage of the head, tail, and ears.

Breed Disease Predilections

Wondering what illnesses to which your beloved cat is prone? Trying to choose a hearty kitten? Take a look at this general list of feline breeds and the health problems to which each is predisposed:

Abyssinian

- Gingivitis.
- Psychogenic alopecia.

American Curl

- Ear infections.

Balinese (Modern/Extreme Wedge-Type Breed Only)

- Easy loss of body heat.
- Heart disease.
- Upper-respiratory infections.

Bengal

- Extreme sensitivities to anesthetics, vaccines, and pesticides.
- Luxating patella (kneecap).

Bombay (Contemporary-Type Breed Only)

- Cherry eye.
- Cleft palate.
- Skull, jaw, and tooth malformations.

British Shorthair

- Heart disease.
- Hemophilia B.
- Luxating patella (kneecap).

Burmese

- Cherry eye (in contemporary-type breed only).
- Cleft palate (in contemporary-type breed only).
- Psychogenic alopecia.
- Skull, jaw, and tooth malformations (in contemporary-type breed only).
- Vestibular disease.

Chantilly/Tiffany

- Psychological stress when left alone frequently.

Chartreux

- Luxating patella (kneecap).

Cornish Rex

- Easy loss of body heat.
- Heart disease.
- Thyroid deficiencies.

Devon Rex

- Easy loss of body heat.
- Heart disease.
- Hip dysplasia.
- Luxating patella (kneecap).
- Spasticity.

Egyptian Mau

- Extreme sensitivities to anesthetics, vaccines, and pesticides.
- Heart disease.
- Luxating patella (kneecap).

Exotic Shorthair

- Epiphora (excessive tearing).
- Sinus conditions.

Havana

- Respiratory infections.

Himalayan

- Epiphora (excessive tearing) (in modern-type breed only).
- Feline hyperesthesia syndrome.
- Psychogenic alopecia.
- Sinus conditions (in modern-type breed only).

Korat

- Extreme sensitivities to anesthetics, vaccines, and pesticides.
- Respiratory infections.

Maine Coon

- Hip dysplasia.

Manx

- Abnormal anal opening.
- Central nervous-system disease.
- Corneal dystrophy.
- Spina bifida.

Norwegian Forest

- Hip dysplasia.

Oriental Shorthair

- Extreme sensitivities to anesthetics, vaccines, and pesticides.
- Heart disease.
- Respiratory infections.

Persian

- Entropion.
- Epiphora (excessive tearing).
- Feline hyperesthesia syndrome.
- Heart disease.
- Retinal degeneration.
- Skin sensitivities.
- Stenosis of nasolacrimal ducts.
- Upper and lower teeth that don't meet (Peke-faced variation only).

Scottish Fold

- Epiphora (excessive tearing).
- Extreme sensitivities to anesthetics, vaccines, and pesticides.
- Osteodystrophy.
- Sinus conditions.

Siamese

- Agenesis (lack of development) of upper eyelid.
- Central nervous-system diseases.
- Feline hyperesthesia syndrome.
- Feline Maratoeaux-Lamy syndrome (bone enlargement of ribs and ends of long bones).
- Heart disease.
- Psychogenic alopecia.
- Respiratory infections.
- Vestibular disease.

Somali

- Gingivitis.

Sphynx

- Easy loss of body heat.
- Extreme sensitivities to anesthetics, vaccines, and pesticides.
- Heart disease.

Tonkinese

- Extreme sensitivities to anesthetics, vaccines, and pesticides.
- Respiratory infections.

Turkish Angora

- Deafness.
- Luxating patella (kneecap).

White Domestic Shorthair/Longhair

- Deafness in cats with blue eyes.
- Skin cancer.

Important Questions to Answer Before Going in for an Exam

Being able to give your veterinarian the answers to the following questions will be of great value in helping him come up with a quick and accurate diagnosis:

1. What is the major health concern that brought you and your pet to the clinic? What other minor concerns do you have?
2. When did you first notice something was wrong?
3. What was the first sign of illness you observed?
4. List, in order of time of appearance, the other symptoms you have noticed. Which of these symptoms are still present? Have they improved, gotten worse, or remained the same?
5. Have you done anything at home to treat the problem?
6. How is the animal's appetite, thirst, urination, defecation, and activity level?
7. Have you noticed any changes in your cat's behavior, movement, or breathing?
8. If there is lameness, be sure you know which leg has been favored at home.
9. Has your pet been to another veterinarian for the same problem? If so, what tests did he perform and what were the results? What was his diagnosis? Was medication dispensed? If so, what kind and did it help?

List of Recommended Dosages

How to Use This Chart

The doses of vitamins, minerals, herbs, and enzymes listed here are in some cases higher than the scientifically determined daily requirements that are recommended for a healthy diet. Many of the vitamin and mineral dosages listed here are megadoses used to support the body in fighting off disease threats and not simply for preventing vitamin and mineral deficiencies. Scientific doses for herbs are not available; therefore, the herb doses recommended here are strictly "anecdotal"—meaning doses that Dr. Simon and other veterinary practitioners have determined from personal clinical experience. Every attempt has been made to be cautious and safe in such recommendations; however, individual pet owners must decide whether to use such dosage information to treat their cats. Although Dr. Simon uses these dosages in his own practice, be aware that individual sensitivities and allergic reactions are always possible. If your pet develops any adverse symptoms shortly after beginning such supplements, discontinue using the supplements and call your veterinarian for advice. The doses provided here are for treating a current condition, and generally are not intended for long-term supplementation.

	Description	Dosages
Activated charcoal suspension (also known as micronized charcoal)	Absorbs toxins; used to treat ingestion of poison	3 to 6 ml per pound given orally; repeat dose in 1 hour
Aloe vera juice	Juice made from the aloe vera plant; can be used internally as a bowel cleanser and protects the intestinal wall lining; as an anti-inflammatory agent, it can help treat gastritis, colitis, and enteritis	A cat under 6 pounds should be given 1 teaspoon twice a day; a cat between 6 and 12 pounds should be given 2 teaspoons twice a day; a cat over 12 pounds should be given 1 tablespoon twice a day
Amino acid preparation	Nutrient or protein supplement comprised of various amino acids	Cats should be given the following fractions of the recommended adult human dose: A cat under 6 pounds should be given 1/30; a cat between 6 and 12 pounds should be given 1/15; a cat over 12 pounds should be given 1/10
Benadryl	Antihistamine	1 to 2 mg per pound given orally every 6 to 8 hours

Where to Find	Additional Info
Ask your pharmacist or veterinarian to order it for you	
Health-food stores	Because most forms of aloe vera juice are tasteless, you may want to mix the juice with your cat's food
Health-food stores and veterinarians' offices	Many vets carry a type called VAL syrup, which is made specifically for pets and contains B vitamins in addition to amino acids
Drugstores	If your cat seems significantly drowsy after taking the medication, decrease the dose. (Benadryl may have a sedative effect.)

	Description	Dosages
Beta-carotene	An excellent antioxidant that is nontoxic	A cat under 6 pounds should be given the equivalent of 2,500 IU of vitamin A daily; a cat between 6 and 12 pounds should be given the equivalent of 5,000 IU of vitamin A daily; a cat over 12 pounds should be given the equivalent of 10,000 IU of vitamin A daily
Chlorpheniramine	Antihistamine	A cat under 6 pounds should be given 2 mg orally every 8 to 12 hours: a cat between 6 and 12 pounds should be given 3 mg orally every 8 to 12 hours; a cat over 12 pounds should be given 4 mg orally every 8 to 12 hours.
Chondroitin sulfate	A naturally occurring compound made up of a combination of protein and carbohydrates; protects joints and can be used to treat arthritis	A cat under 6 pounds should be given 100 mg daily; a cat between 6 and 12 pounds should be given 200 mg daily; a cat over 12 pounds should be given 300 mg daily

Where to Find	Additional Info
Drugstores or health-food stores	Unlike dogs and humans, cats cannot convert beta-carotene to vitamin A, so you must supplement your cat's diet with vitamin A as well
Drugstores or veterinarians' offices	A good brand name to look for is Chlor-Trimeton
Health-food stores, veterinarians' offices, and some drugstores	Often used in combination with glucosamine sulfate; if using both, give your cat half doses of each

	Description	Dosages
Chromium	Essential trace mineral used to treat diabetes	A cat under 6 pounds should be given 3 micrograms daily; a cat between 6 and 12 pounds should be given 10 mcg daily; a cat over 12 pounds should be given 20 mcg daily
Cod liver oil	Used to treat corneal ulcers and/or erosions	1 drop in affected eye daily
Coenzyme Q-10	Essential antioxidant nutrient that helps with gum and heart problems	A cat under 6 pounds should be given 2 mg daily; a cat between 6 and 12 pounds should be given 5 mg daily; a cat over 12 pounds should be given 10 mg daily
Colloidal silver	A suspension of tiny silver particles in water	For topical use on burns and wounds, use as a flushing preparation 3 times a day

Where to Find	Additional Info
Drugstores or health-food stores	
Drugstores	Before treating, check with your vet: corneal ulcers and erosions can be quite dangerous, so it is best to use this treatment with veterinary supervision
Drugstores, health-food stores, and veterinarians' offices	Coenzyme Q-10 is very close to being officially named an essential, fat-soluble vitamin
Health-food stores and veterinarians' offices	

	Description	Dosages
Dimethylglycine	Vitaminlike supplement that can be used as an immune system stimulant	A cat under 6 pounds should be given 10 mg daily; a cat between 6 and 12 pounds should be given 25 mg daily; a cat over 12 pounds should be given 50 mg daily; see additional info. column
Echinacea	Herb used to promote healing of wounds and improve the immune system	Cats should be given the following fractions of the recommended adult human dose: a cat under 6 pounds should be given 1/40; a cat between 6 and 12 pounds should be given 1/16; a cat over 12 pounds should be given 1/8; use at 10-day intervals, separated by a 7-day rest; stop use after 3 10-day trials
Flaxseed oil	Herbal oil that encourages healthy skin and a full hair coat; a natural anti-inflammatory agent and immune-system modulator	Cats should be given the following fractions of the recommended adult human dose: a cat under 6 pounds should be given 1/16; a cat between 6 and 12 pounds should be given 1/8; a cat over 12 pounds should be given 1/4

Where to Find	Additional Info
Health-food stores and veterinarians' office	Because different companies use different concentrations of dimethylglycine in their solutions, you must check the concentration of milligrams per milliliter listed on the bottle, then calculate the number of milliliters to give your cat
Health-food stores	Obtain organic freeze-dried sources when possible
Drugstores or health-food stores	Keep refrigerated so that the oil doesn't become rancid; purchase a human-grade, organic, cold-pressed form of the oil, preferably in gelatin capsules

	Description	Dosages
Gatorade	Sugar and electrolyte drink; can be used to prevent dehydration when treating repetivie vomiting and diarrhea	Put 1/8-inch in bowl in place of water; when cat empties bowl, wait 20 minutes, then place another 1/8-inch in bowl; repeat until vomiting subsides
Glucosamine sulfate	A naturally occurring compound made up of a combination of protein and carbohydrates; protects joints and can be used to treat arthritis	A cat under 6 pounds should be given 100 mg daily; a cat between 6 and 12 pounds should be given 200 mg daily; a cat over 12 pounds should be given 300 mg daily
Goldenseal	Herb often used as a natural antibacterial substance; also has anti-inflammatory properties	Cats should be given the following fractions of the recommended adult human dose: A cat under 6 pounds should be given 1/20; a cat between 6 and 12 pounds should be given 1/10; a cat over 12 pounds should be given 1/5
Kaopectate	Absorbent used to treat diarrhea and vomiting	0.5 to 1.0 ml per pound given orally every 2 to 6 hours

Where to Find	Additional Info
Grocery store	
Health-food stores, veterinarians' offices, and some drugstores	Often used in combination with chondroitin sulfate; if using both, give the cat half doses of each
Health-food stores	
Drugstores	

	Description	**Dosages**
Lactobacillus	The "good" bacteria naturally present in the intestines of healthy animals that controls the "bad" bacteria and yeast; synthesizes B vitamins and provides the cells of the intestinal lining with fatty acids	Cats should be given the following fractions of the recommended adult human dose: A cat under 6 pounds should be given 1/16; a cat between 6 and 12 pounds should be given 1/8; a cat over 12 pounds should be given 1/4
Metamucil (Psyllium husks, not seed)	Natural source of fiber that acts as a bulk cathartic and prevents and treats constipation	Cats should be given the following fractions of the recommended adult human dose: A cat under 6 pounds should be given 1/20; a cat between 6 and 12 pounds should be given 1/10; a cat over 12 pounds should be given 1/5

Where to Find	Additional Info
Drugstores, health-food stores, and veterinarians' offices	When purchasing, opt for high-quality brand names found in a refrigerated area of the store to ensure that the viability of the live bacteria is maintained
Drugstores, health-food stores, and veterinarians' offices	Increasing your pet's water consumption is very important when supplying psyllium; if no bowel movement is produced in 48 hours, see your veterinarian

	Description	Dosages
Plant-derived digestive-enzyme supplements	A source of digestive enzymes that helps the body to digest its food	Cats should be given the following fractions of the recommended adult human dose: A cat under 6 pounds should be given 1/16; a cat between 6 and 12 pounds should be given 1/8; a cat over 12 pounds should be given 1/4; dose should be sprinkled over lightly dampened food 10 minutes before it is served
Proteolytic enzyme supplement	A specific type of digestive-enzyme supplement that contains only protease; helps to strengthen the immune system	Cats should be given the following fractions of the recommended adult human dose: A cat under 6 pounds should be given 1/16; a cat between 6 and 12 pounds should be given 1/8; a cat over 12 pounds should be given 1/4

Where to Find	Additional Info
Health-food stores	Buy a brand name for humans unless a pet supplement is available; the supplement should contain amylase, protease, lipase, and cellulase
Health-food stores	For more effective results, give apart from meals; bromelain or papain tablets are recommended

	Description	Dosages
Sulfur	Mineral required for synthesis of body proteins; can also act as an antioxidant	A cat under 6 pounds should be given 50 mg in the form of methylsulfonyl methane daily; a cat between 6 and 12 pounds should be given 150 mg daily; a cat over 12 pounds should be given 250 mg daily
Taurine	Essential amino acid used as a nutritional supplement to help with certain heart problems and seizures	A cat under 6 pounds should be given 125 mg twice daily; a cat between 6 and 12 pounds should be given 250 mg twice daily; a cat over 12 pounds should be given 500 mg twice daily

Where to Find	Additional Info
Health-food stores	Purchase in form of methylsulfonyl methane
Health-food stores and pet stores	

	Description	Dosages
Trace mineral supplements	Essential minerals that keep the body functioning properly	Depending on weight, cats should be given the following fractions of the recommended adult human dose: A cat under 6 pounds should be given 1/20; a cat between 6 and 12 pounds should be given 1/8; a cat over 12 pounds should be given 1/4
Vitamin A	Antioxidant and immune-system stimulant; as a fat-soluble vitamin it is toxic if dose is too high	A cat under 6 pounds should be given 1,000 IU daily; a cat between 6 and 12 pounds should be given 3,000 IU daily; a cat over 12 pounds should be given 6,000 IU daily
Vitamin B complex	Vitamin that helps to maintain healthy nerves, skin, eyes, hair, liver, and mouth	Follow dosage recommended on a bottle of a brand made especially for pets

Where to Find	Additional Info
Health-food stores	Best form to purchase is chelated trace mineral tablets that contain as many as 74 different trace minerals
Drugstores and health-food stores	Because of its potential toxicity, only use for short periods (no more than 2 weeks) without stopping; best when used with a veterinarian's supervision
Pet stores or veterinarians' offices	Your vet can recommend a brand made specifically for pets

	Description	Dosages
Vitamin B$_6$	Nontoxic, water-soluble B vitamin; brewer's yeast is an excellent source of B$_6$	In the form of brewer's yeast, a cat under 6 pounds should be given 1/4 teaspoon daily; a cat between 6 and 12 pounds should be given 1/2 teaspoon daily; a cat over 12 pounds should be given 3/4 teaspoon daily
Vitamin B$_{12}$	Nontoxic, water-soluble B vitamin; works well as an appetite stimulant and energy booster; can be helpful in treating anemia	You can provide naturally by supplementing your cat's food with organ meat (kidney, liver, heart, etc.) served raw or lightly cooked
Vitamin C	Vitamin used for strengthening the immune system and as an anti-allergic, anti-inflammatory, anti-bacterial, antiviral, and detoxicant agent	A cat under 6 pounds should be given 100 mg daily; a cat between 6 and 12 pounds should be given 250 mg daily; a cat over 12 pounds should be given 500 mg daily; reduce the dose if a soft stool develops

Where to Find	Additional Info
Drugstores and health-food stores	If you do not supply in brewer's yeast form, you can provide the necessary amount by giving your cat a multivitamin/mineral supplement
Meat available from any butcher or grocery store; multivitamin/mineral supplement available from veterinarians' offices or pet stores	Your vet may be able to recommend a multi-vitamin/mineral supplement made specifically for cats that includes the necessary amount of B_{12}
Drugstores or health-food stores	Best purchased in calcium ascorbate or sodium ascorbate forms rather than acidic acid form, which can upset the stomach; try to obtain a brand that includes bioflavonoids; if you are using vitamin C to acidify the urine, choose the ascorbic acid form

	Description	**Dosages**
Vitamin E	Essential antioxidant vitamin used for strengthening the immune system	A cat under 6 pounds should be given 50 IU daily; a cat between 6 and 12 pounds should be given 100 IU daily; a cat over 12 pounds should be given 200 IU daily
Zinc	Essential mineral used for tissue repair and healing, proper immune-system functioning, and healthy skin and coat	A cat under 6 pounds should be given 2 mg daily; a cat between 6 and 12 pounds should be given 5 mg daily; a cat over 12 pounds should be given 8 mg daily

Where to Find	Additional Info
Drugstore or health-food stores	Because it is a fat-soluble substance that accumulates in the body, it can become toxic in high doses
Drugstores, health-food stores, or veterinarians' offices	Absorbed most efficiently if purchased in its chelated form; potentially toxic in high doses, so be sure you are not supplying it in any other supplements; take with a copper supplement, because it may interfere with absorption of naturally occurring copper

Index

About the Authors

JOHN M. SIMON, DVM, owns a private practice, Woodside Animal Hospital, in Royal Oak, Michigan. A graduate of the Michigan State University School of Veterinary Medicine, Dr. Simon has over thirty years of experience in conventional and alternative pet care. In 1982, Dr. Simon became Detroit's first certified veterinary acupuncturist; in 1996, he received his certificate from the American Veterinary Chiropractic Association. Dr. Simon is a past president of the Oakland County Veterinary Medical Association (OCVMA) and has served as both an officer and board member of the Southeastern Michigan Veterinary Medical Association (SEMVMA). In 1993, he began hosting his own weekly cable TV talk show entitled, *Your Pet's Good Health.* Author of *Basic Bird Care & Preventive Medicine,* Dr. Simon has written a regular column in Detroit's *Daily Tribune* since 1983 and has contributed extensively to pet-care magazines such as *Natural Pet.* His practice has been featured on both the local and national television news. Dr. Simon lives in Franklin Village, Michigan, with his wife and two children.

STEPHANIE PEDERSEN is a freelance writer and editor who specializes in the areas of health and beauty. Her articles have appeared in numerous publications, including *American Woman, Sassy, Teen, Unique Homes, Weight Watchers,* and *Woman's World.* She has also cowritten *What Your Dog Is Trying to Tell You: A Head-to-Tail Guide to Your Dog's Symptoms— and Their Solutions* with Dr. Simon. She currently resides in New York City.

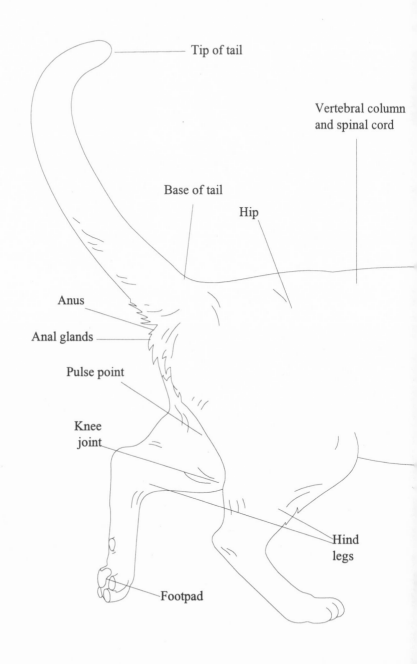

Tip of tail

Vertebral column
and spinal cord

Base of tail

Hip

Anus

Anal glands

Pulse point

Knee
joint

Hind
legs

Footpad